S0-BCY-968

INFORMATION
TECHNOLOGY
ON THE MOVE

WILEY SERIES IN COMMUNICATION AND DISTRIBUTED SYSTEMS

Editorial Advisory Board

Professor B. Evans
University of Surrey

Professor G. Pujolle
Université Pierre et Marie Curie

Professor A. Danthine
Université de Liège

Professor O. Spaniol
Technical University of Aachen

Integrated Digital Communications Networks (Volume 1 and 2)
G. Pujolle, D. Seret, D. Dromard and E. Horlait

Security for Computer Networks, Second Edition
D.W. Davies and W.L. Price

Elements of Digital Communication
J.C. Bic, D. Dupontiel and J.C. Imbeaux

Satellite Communications Systems, Second Edition (Systems, Techniques and Technology)
G. Maral and M. Bousquet

Using Formal Description Techniques (An Introduction to ESTELLE, LOTOS and SDL)
Edited by Kenneth J. Turner

Security Architecture for Open Distributed Systems
S. Muftic, A. Patel, P. Sanders, R. Colon, J. Heijnsdijk and U. Pulkkinen

Future Trends in Telecommunications
R.J. Horrocks and R.W.A. Scarr

Mobile Communications
A. Jagoda and M. de Villepin

Information Technology on the Move (Technical and Behavioural Evaluations of Mobile Telecommunications)
G. Underwood, F. Sommerville, J.D.M. Underwood and W. Hengeveld

INFORMATION TECHNOLOGY ON THE MOVE

Technical and Behavioural Evaluations of Mobile Telecommunications

Geoffrey Underwood
University of Nottingham
UK

Jean D. M. Underwood
University of Leicester
UK

Fraser Sommerville
Castle Rock Consultants,
Nottingham, UK

Willem Hengeveld
Bakkenist Management Consultants,
Amsterdam, The Netherlands

JOHN WILEY & SONS
Chichester • New York • Brisbane • Toronto • Singapore

621.3845

I 43

Copyright © 1994 by John Wiley & Sons Ltd,
Baffins Lane, Chichester,
West Sussex PO19 1UD, England

All rights reserved.

No part of this book may be reproduced by any means,
or transmitted, or translated into a machine language
without the written permission of the publisher.

Other Wiley Editorial Offices

John Wiley & Sons, Inc., 605 Third Avenue,
New York, NY 10158-0012, USA

Jacaranda Wiley Ltd, G.P.O. Box 859, Brisbane,
Queensland 4001, Australia

John Wiley & Sons (Canada) Ltd, 22 Worcester Road,
Rexdale, Ontario M9W 1L1, Canada

John Wiley & Sons (SEA) Pte Ltd, 37 Jalan Pemimpin #05-04,
Block B, Union Industrial Building, Singapore 2057

Library of Congress Cataloging-in-Publication Data

Underwood, Geoffrey (Geoffrey D. M.)
    Information technology on the move : technical and behavioural
    evaluations of mobile telecommunications / Geoffrey Underwood ...
    [et al.].
        p.   cm.
    Includes bibliographical references and index.
    ISBN 0 471 93850 5
        1. Intelligent Vehicle Highway Systems—European Economic
    Community countries.   2. Transportation, Automotive—European
    Economic Community countries—Communication systems—Technological
    innovations.   3. Automobile driving—European Economic Community
    countries.   I. Title.
    TE228.3.U53   1993
    388.3' 124' 094—dc20   93-10043
                                                                   CIP

British Library Cataloguing in Publication Data

A catalogue record for this book is available from the British Library

ISBN 0 471 93850 5

Typeset in 10/12pt Palatino from author's disks by Text Processing Department,
John Wiley & Sons Ltd, Chichester
Printed and bound in Great Britain by Biddles Ltd, Guildford, Surrey

CONTENTS

PREFACE

Our roads are overcrowded, inefficient, environmentally damaging, and dangerous to all who use them. One solution to the crowding problem is to build even more roads, but this is clearly unacceptable as a long-term solution—apart from limits to the available space and resources there are environmental reasons for arguing against continuous expansion. Local and national authorities have implemented traffic management schemes but the widespread availability of information technology now offers us a range of new solutions. These solutions involve what is known collectively in Europe as road transport informatics, or RTI, led by an initiative of the European Community using the acronym of DRIVE (Dedicated Road Infrastructure for Vehicle Safety in Europe). This is not an exclusively European venture, of course, and related systems are being developed in Japan under the VICS programme (Vehicle Information and Communication Systems) and are known in the United States as Intelligent Vehicle Highway Systems (IVHS). By developing road infrastructures and vehicles that can communicate directly, an integrated environment can be developed in which vehicles will be seen to co-operate with each other and with the road system with the result that safety, efficiency, and environmental quality will all be improved.

The fully integrated RTI solution is in the future, of course, but already products are available which point to the way ahead. In-vehicle navigation and radio message systems, for example, are able to provide the driver with route guidance information, and as well as satisfying the individual user's needs to reduce journey time they serve the function of alleviating congestion by allowing that vehicle to occupy the road for less time. *Information Technology on the Move* provides an introduction to the systems currently available and under development, and describes a procedure whereby potential RTI systems can be matched against need. The primary criteria in this assessment involve the technical characteristics of the proposed product and the match between the system and need, but one of our aims has been to point to the importance of user-centred design. The system under development should be seen as an interactive system in which the human user is a component with characteristics that, if neglected or denied, will certainly result in inefficient use or even non-use of the technology. This concern for user-centred design is of greater importance for certain RTI systems than for others, of course—with the design of in-vehicle navigation displays and with co-operative driving systems, rather than with centralised traffic management systems, perhaps—but whenever humans interact with technology it is vital for the configuration of both human and computer abilities to be taken into account.

Accordingly this is a guide to the potential of information technology and a guide to the assessment of systems under development. It has its origins in the DRIVE programme, at a point when participants were expanding their development and testing of new systems. One problem encountered at this stage was that progress was slower than it might have been because the engineers, economists, geographers, psychologists and computer scientists who were collaborating were learning about the systems as they were being developed. There was no bank of trained research and development staff available to work on the projects for the very reason that this was new ground. As part of the response to this recognised skill shortage the European COMETT programme, which itself is concerned with training in a wide range of technological fields, established a project to make available information about the development and testing of RTI systems. The partners who worked on this collaborative COMETT project are the authors of *Information Technology on the Move*, which is one of the products of the project.

Although four members of the project team are responsible for the book, others should be acknowledged for their valued contributions at different stages. Principal among these is Peter Davies of Castle Rock Consultants, who was fundamental in seeing the possibilities available through linking the DRIVE and COMETT programmes. Peter's contribution was as originator, and his perceptions were essential in bringing the project together. Other colleagues in Amsterdam and Nottingham who helped us make progress are Leo van der Hart, Piet van Berkel, Grant Klein, Hugh Milton, Sara Willott, Mark Leicester, Mark Torrance, and Sue Jeffery.

We have included some material that has been published elsewhere or is currently available only from technical reports, and thanks are due to the original authors and publishers who have given permission for this reproduction. In particular we acknowledge Siemens-Plessey (for Figure 1.4), Taylor and Francis Ltd (for material on behavioural adaptation in Chapter 2), Artech House (for material on mobile information systems in Chapter 6), and the Transportation and Traffic Research Division of Rijkswaterstaat (for material from the case study on Amsterdam in Chapter 7).

Our aim has been to describe some of the uses of information technology in improving our use of the road network. We hope that it will be of interest to students and engineers involved in the development of road and vehicle RTI systems, as well as to members of disciplines who are co-operating in attempts to solve the problem of congested, unsafe and polluted roads by using these systems.

Geoffrey Underwood,
University of Nottingham

Fraser Sommerville,
Castle Rock Consultants, Nottingham

Jean D. M. Underwood,
University of Leicester

Willem Hengeveld,
Bakkenist Management Consultants, Amsterdam

Section 1

AN INTRODUCTION TO ROAD TRANSPORT INFORMATICS

1 ROAD TRANSPORT INFORMATICS: THE APPLICATION OF INFORMATION TECHNOLOGY TO TRAFFIC PROBLEMS

AN OVERVIEW OF THE PROBLEM DOMAIN

Road transport in the developed and developing world is an urgent and tangible problem impinging on our society as a whole and the lives of each of us as individuals. Anyone who regularly uses Europe's roads will be aware that traffic congestion is often the normal state of affairs rather than an occasional inconvenience. The most obvious effect of this congestion is in increased journey times for both the private and the commercial motorist. These increased journey times are serious both in social and economic terms but equally significant issues are found in the escalating traffic accident rates and the levels of environmental damage that result from an overloaded traffic system. These interrelated issues of road transport were identified as key areas of concern under the European Commission's DRIVE research programme (Dedicated Road Infrastructure for Vehicle Safety in Europe), and by similar large-scale programmes in the United States and in Japan. The common goal of these initiatives is the development of systems that use new technologies in the solution of traffic problems.

The specific objectives of DRIVE are to improve road safety, to maximise road transport efficiency and to contribute to environmental improvements. In this book we shall describe some of the research conducted under the umbrella of the DRIVE programme, focusing upon the transmission of information between the road infrastructure and the vehicles using the road network. These are systems that can provide, for example, current route guidance and parking information to drivers, and can be used for the automatic collection of tolls. Information collected by the infrastructure can monitor traffic flow, detect incidents requiring special attention, and be used by automated traffic control systems. One of the important benefits of the use of technology here is the immediacy of the information. The driver can collect route guidance advice that is sensitive to traffic congestion as it is accumulating, and a traffic control system can also respond to such congestion in real time. In *Information Technology on the Move* we shall also be focusing upon the assessment of systems proposed for the purpose of transmitting information between vehicle and infrastructure. The assessment procedure will mainly involve the technical requirements of the system, but will consider the solution of the traffic problem from the perspective of the operator of the system—the driver.

Solving transport problems with information technology

How do we alleviate the interrelated set of transport problems that prompted the DRIVE initiative? Quite simply, as with all demand–supply problems, solutions may be viewed as one of increasing capacity to meet demand or damping down demand to a level deliverable under the current supply conditions. As there appears to be no finite limit to road traffic demand, as recent history has shown, increased capacity is no solution to the problem of supply and it does little if anything to resolve the environmental problems. Reducing demand to fit the existing infrastructure would appear a cheap environmentally friendly solution but it may seriously affect the mobility of the individual and commercial operations. Both solutions may prove costly and a demand reduction solution, as it impinges on personal freedoms, may require politically sensitive legislative change.

An intermediate, partial solution, would be to make more efficient and effective use of existing road networks. Efficiency gains through the increased use of public transport brought about by fare subsidies and/or the reduction of journey times through the use of dedicated bus lanes are widely used solutions throughout Europe. Encouraging drivers to give a lift to a friend ("car pooling"), through the designation of priority lanes, may also help to alleviate the problems associated with peak traffic flow.

These solutions have proved to be a significant but not sufficient solution to traffic problems. Other solutions that involve the use of information technology are being sought and this is the overall goal of the research into Road Transport Informatics (RTI)—sometimes referred to as Advanced Road Transport Telematics (ATT) in Europe, as Intelligent Vehicle Highway Systems (IVHS) in the United States, and as Vehicle Information and Communication Systems (VICS) in Japan. These descriptors and acronyms will be considered to be synonymous throughout our discussions. Even though different descriptors are preferred in different research programmes they have the common feature of investigating the use of information technology to improve traffic systems. This research also endeavours to find additional, technologically supported efficiency gains by influencing the day-to-day behaviour of vehicle drivers. We will return to behavioural issues relating to the implementation of RTI and other road traffic solutions—particularly behavioural adaptations to engineering innovation—in the next chapter. The need to recognise human sensitivities that positively or negatively affect the success of the RTI implementation process are underestimated at the planner's and engineer's peril.

The European Commission has indicated both its recognition of the complex supply–demand problem and its preference for solutions based on efficiency gains where possible, by its instigation of and support for the DRIVE research programme. Similar initiatives are under way world wide, for example, the VICS programme in Japan and the IVHS programme in the United States. Most of the examples of RTI research and development to be described in this book are taken from the European DRIVE programme, but it is worth while to demonstrate the more general interest and involvement with the use of information technology to solve road transport problems by mentioning a few US and Japanese programmes.

Road transport informatics in the United States

Work on RTI in the United States has been brought within the IVHS-America framework. IVHS (Intelligent Vehicle Highway System) encompasses similar technologies and applications to the European DRIVE initiative, and this work is now co-ordinated by the Intelligent Vehicle Highway Society of America (see Shields, 1991). This is a non-profit educational and scientific association established to co-ordinate and foster public and private partnerships that will use information technology to contribute to an improved US transportation system. Within the IVHS programme there are a number of large-scale demonstrations of RTI technologies planned or under way, and some of these will be described briefly here.

The aim of the SMART Corridor study is to integrate traffic sensors, computers and communications links on a 15-mile section of the Santa Monica Freeway and five adjacent arterials. This is a co-operative demonstration project currently being implemented by a number of local and Federal agencies in the Los Angeles area. The system will provide information on current traffic conditions that will be used to assist highway agencies in making control decisions. In addition the system will provide up-to-date information to motorists through a variety of media. The SMART Corridor project is funded at a level of approximately $50 million, and began in 1989 with a conceptual design study. One of the major focuses of the study concerns the methods by which information can be communicated to and from motorists in the corridor. A number of alternative approaches are being evaluated, including variable message signs, highway advisory radio, cellular telephone systems, teletext and videotext.

Closely related to the SMART Corridor study is the PATHFINDER project.

This is a co-operative field experiment involving the California Department of Transportation (Caltrans), the Federal Highways Authority (FHWA) and the General Motors vehicle manufacturer, which aims to perform an initial assessment of the feasibility and utility of a real-time in-vehicle navigation and motorist information system. The project is being tested within the SMART Corridor demonstration area.

The in-vehicle equipment selected for Pathfinder is the Bosch TravelPilot version of the ETAK Navigator. In addition to conventional map display facilities, the system includes a special two-way, time-multiplexed communications element. This is used to send vehicle speed and location data to a control centre, and to return pertinent traffic data to the in-vehicle unit. Areas of congestion are identified on the map display, allowing the driver to select an alternative route if necessary.

Testing of the Pathfinder system began in June 1990, using Caltrans employees, to evaluate the ability of the system to avoid congested areas. In a second phase of testing, hired drivers were used to establish time savings achieved through use of the system between various origins and destinations. A final test phase examined the acceptability and perceived utility of the system for commercial drivers.

TRAVTEK is a co-operative project currently being undertaken in Orlando, Florida, by the FHWA, General Motors, the American Automobile Association, Florida Department of Transportation and the City of Orlando. The study has demonstrated the use of prototype in-vehicle equipment for the provision of up-to-date traffic condition, routing and tourist information. TRAVTEK costs approximately $8 million and involves 75 GM rental cars and 25 vehicles used by high-mileage local drivers.

The test vehicles are equipped with navigation and driver information systems linked to information centres by radio data communications. The in-vehicle video monitor is capable of displaying maps of the Orlando area, including areas of congestion and services, text information on traffic incidents or services, and route guidance instructions using simple graphical cues. A traffic management centre is responsible for combining and transmitting traffic information received from a variety of sources.

A related co-operative project is ADVANCE. This is an operational test of a dynamic route guidance system to be undertaken in the Chicago suburbs. Project partners include the FHWA, Illinois DOT, Illinois Universities Transportation Research Consortium, and Motorola. The project began in 1990 with an estimated budget of over $40 million over five years. ADVANCE utilises a Motorola in-vehicle navigation system as the basis of presenting real-time routing instructions covering both freeways and arterial streets. Traffic information is collected using both conventional sensors and the vehicles themselves acting as probes. In this latter case, data are transmitted to a traffic information centre via RF transmissions or global positioning system (GPS) satellites.

Another organisation undertaking a major IVHS initiative is TRANSCOM (the Transportation Operations Co-ordinating Committee). This is a consortium of transportation agencies in the New York/New Jersey metropolitan area. TRANSCOM is responsible for co-ordinating regional traffic management through increased inter-agency co-operation, review of procedures and programmes, formulation of mutual assistance programmes and co-ordination of planned construction and maintenance activities.

The TRANSCOM IVHS study investigates the use of automatic vehicle identification (AVI) technology for traffic monitoring and management. AVI equipment is already being used in the New York/New Jersey area for electronic toll and traffic management on facilities operated by the Port Authority of New York and New Jersey and the Triborough Bridge and Tunnel Authority. The TRANSCOM study will therefore aim to use the same AVI technology in an integrated system with subsequent cost-sharing benefits.

The first stage of the TRANSCOM study focused on the design of an area-wide traffic management system. This examined issues such as the required density of AVI readers, the number of equipped vehicles, and communications and software needs. After general installation of equipment in the New York/New Jersey area, TRANSCOM evaluated system performance from the perspectives of traffic surveillance facilities, incident detection and response capabilities, and clarity of data. TRANSCOM also examines the cost-sharing options for the system and the market for traffic data collected by the AVI technology.

The ENTERPRISE programme (Evaluating New Technologies for Roads Program Initiatives in Safety and Efficiency) is a multi-state IVHS initiative for co-operative IVHS developments within a co-ordinated framework. The overall theme of ENTERPRISE is the development and demonstration of digital communication technologies. ENTERPRISE operates as an umbrella programme, under which each participating member will address distinct projects and areas of IVHS. Guidestar is one of the key programmes under the umbrella, along with Arizona's MIDAS programme and the programmes in Iowa, Colorado, Michigan, Washington and Ontario. Overall, the principal objectives of the ENTERPRISE initiative can be summarised as:

- to establish a group of state, federal and provincial bodies with similar goals and objectives and a feeling of mutual trust;

- to facilitate each participant in selecting and conducting its own IVHS activities in accordance with specific needs and interests, and in agreement with common programme goals;

- to make rapid progress in the development of IVHS technologies, particularly those concerned with digital communications approaches;

- to operate with a small, tightly controlled management team responsible for co-ordinating the programme in an effective manner;

- to provide increased opportunities for communication and co-operation between participating organisations;

- to create an efficient mechanism for disseminating the results of the programme to higher bodies such as IVHS America, and broadly throughout the IVHS community;

- to achieve true public/private sector co-operation, providing cost saving benefits and ensuring economically sound technology development;

- to make optimum use of funding sources by avoiding duplicative efforts and focusing on carefully targeted areas; and

- to utilise programme synergy, to increase the benefits of IVHS initiatives for programme participants, and thus to make a greater overall contribution to the advancement of IVHS.

PATH (Program on Advanced Technology for the Highway) is a key element of the California Department of Transportation's new technology development programme. The initiative has a proposed budget of $56 million over a six-year period, with sponsorship from Caltrans, the FHWA and the National Highway Traffic Safety Administration. PATH research is conducted primarily by the University of California's Institute of Transportation Studies (ITS) at Berkeley.

The main focus of PATH is vehicle and highway automation, with the ultimate goal of completely automated vehicle operations. A major element of the research is investigation of roadway electrification; this is a continuation of work previously undertaken in the Santa Barbara Electric Bus Programme. PATH is also conducting research into longitudinal and lateral control of vehicles based on a number of alternative technical approaches. In addition, PATH is performing research on traffic management and driver information systems, including consideration of driver behaviour and safety aspects.

Road transport informatics in Japan

Japanese researchers have been investigating a range of advanced transportation technologies since the 1960s. Early work examined traffic management systems, prior to a shift of emphasis toward development of driver information technologies. More recently, automated vehicle control projects have been undertaken in Japan. The main Japanese RTI initiatives currently being carried out or proposed include RACS, AMTICS, VICS AND SSVS.

RACS (Road-Automobile Communication System) has involved the development of a short-range, high data rate microwave communications system. This is used for a number of road-vehicle communication applications including beacon transmissions for the correction of positional errors in vehicle location systems; the provision of a communications link for an externally linked route guidance system; and the ability to pass individual messages or data for display on in-vehicle units.

AMTICS (Advanced Mobile Traffic Information and Communication System), is an integrated traffic information and navigation system, being developed under the guidance of the National Police Agency. The system will display on screen in each vehicle traffic information gathered at traffic control and surveillance centres managed by the police in 74 cities. The major benefit of AMTICS will be its ability to display in real time, not only the vehicle's current position and route, but also information on traffic congestion, regulations, roadworks and parking.

Both AMTICS and RACS are aiming to provide traffic information to drivers. The main technical difference between these two approaches is the communications link. Since their aims are both the same, it was recommended that the two programmes be co-ordinated in some way. The Ministry of Construction, the National Police Agency and the Ministry of Posts and Telecommunications have jointly developed an initiative known as VICS (Vehicle Information and Communication System) to achieve this co-ordination. VICS has set a policy on the future use of mobile communications in Japan covering digital cellular radio, FM subcarrier broadcasting, beacons and package media.

Under the VICS framework, the Ministry of Construction is continuing the work of RACS. This will result in systems for one- and two-way communications between roadside beacons and vehicles. Similarly, AMTICS developments will continue to be pursued under VICS, further investigating the use of broadcasting an FM subcarrier system, similar to the Radio Data System. Also, the existing AMTICS demonstration area is being enlarged.

One of the major RTI programmes which has recently started in Japan is the SSVS (Super Smart Vehicle System) initiative. This aims to promote the research and development of RTI technologies, particularly those concerned with highway safety. The initiative is targeted towards technologies for implementation in the highway environment within the next 20 to 30 years. SSVS addresses a number of factors, including the following:

- technical, social and economic issues concerned with the development and introduction of RTI technologies;

- evaluations of the needs of RTI in terms of safety, efficiency and convenience;

- analysis of the fundamental characteristics of in-vehicle units based on technological capabilities;

- investigation of man–machine interfaces and human factors issues;

- policies concerning technology research, development and evaluation.

The SSVS initiative started in August 1990, and its first two years concentrated on preliminary research. During this period, the programme investigated future socio-economic conditions and vehicle requirements, as well as planning research and development projects and related activities.

RTI APPLICATIONS AND SYSTEMS

Traditionally, the decisions that drivers make about which roads to use and which to avoid during their journeys are influenced by the purpose of their journey, by their personal knowledge of the locality, by roadside signs, and/or by road travel information broadcasts. Roadside signs have a limited capacity to convey information about traffic constrictions, other than by indicating potential blockages such as long-term road works. Information about the day-to-day or hour-to-hour traffic situation is not available through static signalling systems, although variable message systems with real-time data links could overcome this problem of immediacy. Radio broadcasts directly to drivers have the potential to provide information about current traffic conditions but only the largest and most affluent of European and American cities achieve the street-by-street coverage needed to make these systems really effective. The majority of broadcasts cover large areas and focus on key roads and, as they are transmitted at intervals as part of scheduled programming, their usefulness tends to be limited to long-haul journeys on the primary road network.

Effective road traffic management requires the gathering and dissemination of 'real-time' traffic information that is current, accurate and specific to the needs of the individual driver. Road transport informatics, by applying developments in information technology to the problem of road traffic management, makes the integrated and automated handling of traffic information possible. An example of one such system is ALI (Autofahrer Leit und Informationssystem) an externally linked route guidance system developed by Blaupunkt which uses inductive loops for vehicle–roadside communication. This early system has been developed and tested in a number of forms, and can be used here as an example of a popular solution to the guidance problem, the problem of how to deliver information to the driver which is relevant to his or her needs.

ALI: an example of an RTI system

ALI, developed by Blaupunkt, was an RTI system designed to provide up-to-date route advice to drivers. It was designed principally for use on the autobahn. Drivers used a small key pad within their vehicle to enter a seven-digit code to indicate the drivers' intended destinations. Each time the vehicle crossed an autobahn junction this code was transmitted to a roadside receiver and from there the information was passed to a central computer. The central computer used this data to build up an overall picture of current traffic conditions and, using predictive algorithms, forecasted traffic developments. On the basis of these data, routing advice was calculated and passed back to the individual vehicles. The advice was displayed within the vehicle on a small screen. The effectiveness of this system was dependent on a number of factors, including the number of vehicles providing data input and the accuracy to which the proportion of users of the system was known.

The potential benefits of RTI systems lie in their ability to deliver up-to-the-minute information to individual users. This information can be delivered in 'real time', in comparison to road signs and radio broadcasts, which deliver 'old' information. Potential difficulties with the systems are less technical than social and economic. These systems increase in efficiency with increasing consumer involvement, but initiation of such systems is costly and consumers need to be convinced of the benefits before accepting the product. Road traffic signs and radio broadcasts are cheap systems which are 'universally' available with indirect costs to the user hidden in road taxes, the purchase of a car or a broadcast licence fee.

Application areas

While the majority of applications in RTI have focused on traffic management and driver information systems like ALI, information technology as such has a wide potential of applications in the road transport field. Seven application areas can be identified:

- demand management;
- traffic and travel information;
- integrated urban traffic management;
- integrated inter-urban traffic management;
- driver assistance and co-operative driving;
- freight and fleet management;
- public transport management.

RTI implementations are often effective in several of these problem domains, and lead to integrated systems.

Collecting money to pay tolls on roads or bridges can be a major cause of delay as vehicles stop at toll booths. Automatic toll debiting on bridges, tunnels or motorways, in which the toll is automatically recorded and later debited from the driver's bank account, provides better throughput for all classes of motorist and can be seen as part of an overall traffic management plan. The process can be speeded up considerably by electronic communication between vehicles and the toll booth. Warning transmissions similarly meet multiple needs. At their simplest, warning transmission systems relay information from the police and other sources to drivers. They achieve the same goal as traffic bulletins broadcast during regular radio schedules. An improvement on this procedure is the German ARI (Autofahrer Rundfunk Information) system which transmits road traffic information independently of regular radio broadcasts to car radios modified to receive these messages (see Chapter 6). The ALI system is one system designed to address management issues. Simpler systems are possible, such as information about road traffic conditions being relayed to displays at the roadside rather than directly to vehicles.

Autonomous routing can potentially aid both the commercial and private motorist by efficiently guiding drivers to their destinations. Such RTI systems do not in principle contribute to any overall traffic management plan, however, but parking management systems do. These relatively straightforward systems communicate the parking space available in an urban area or airport as vehicles enter that area, reducing both the time and distance travelled by motorists to find a parking space.

One of the key developments for freight and fleet management has been the use of RTI to co-ordinate commercial vehicle movements. This requires both complex route calculations to minimise journey times, and efficient communication between vehicles and their central base. PRODAT, developed by Siemens, permits vehicles spread over a wide area to communicate to a central processor via a satellite receiver. Messages are entered into the system using a keyboard, and returning messages are printed out on a small printer.

Other RTI applications which have been suggested or implemented include automatic vehicle identification for payment, or regulation and enforcement purposes, automatic vehicle location, and driver status monitoring. Here we will focus on those applications that have significant implications for road traffic management.

Implementations of RTI systems

Some RTI systems are designed to meet more than one application. A good example of such a system is ALI-SCOUT, recently developed and re-named EURO-SCOUT. This system combines route guidance with other traffic information including parking, public transport, and road tolls.

Bosch, Blaupunkt and Siemens have extended the basic data captured by ALI with the addition of data mapping of dynamic changes such as roadworks or accidents. Such data do not appear on the base map stored by the central computer. As for ALI, infra-red receiver/transmitter units (or beacons) placed at the road-side record the identity and intended destination of passing vehicles. Strategic placement of such units around a road network facilitates the collection of information about current traffic conditions based on travel time between beacons. This allows predictive forecasts to be made about possible delays. This information, augmented by hazard warnings from the police and other organisations, forms the basis for the route guidance advice which is transmitted back to the vehicles via infra-red beacons.

The ALI-SCOUT system was designed to fulfil several application needs. Its effectiveness depends on the number of data points collected and therefore on the number of vehicles capable of inputting to and receiving output from the system. If enough cars were to be equipped with the necessary technology it would be an effective form of traffic control and in addition it would provide route guidance and hazard warnings. An elaborated system could provide parking information and could be used for automatic debiting purposes. The long-term goal of developments in RTI is the development of an Integrated Road Transport Environment (IRTE). Within an IRTE there should be sufficient information about traffic conditions to make it possible to manage traffic in a way that makes optimal use of the road space available. This proactive approach contrasts sharply with the crisis management that is often currently employed to control traffic in the absence of RTI systems.

Driver information systems provide drivers with information on roadway conditions and route availability. This information increases the efficiency of drivers in carrying out the three key trip activities of route planning, route following and trip chain sequencing. This latter problem of complex trips which have multiple stops or purposes , and which is known as the 'travelling salesman problem', has considerable economic significance. Misplanned or inefficient routing results in lost time, wear and tear on vehicles and direct increases in on-road costs.

A number of RTI implementations supporting the real need to increase trip efficiency are now available. These include:

Electronic route planning: Here systems link minimum path computer algorithms to highway network data bases. They supersede the historical route data provided by maps or memory. Recent examples include the pan-European "ATIS", "ROADWATCH" developed by the AA in the UK, and the French "ANTIOPE".

Traffic information broadcasting systems: Here motorists are provided with regular updates of current road traffic conditions, enabling them to adapt and re-plan their route as necessary. Various conditions can alter the traffic situation, such as recurring congestion, road construction, road maintenance, or short-term hold-ups caused by a traffic incident. They achieve the same goal as traffic bulletins broadcast during regular radio schedules. The German ARI system transmits road traffic information, independently of regular radio broadcasts, to car radios modified to receive these messages. Drivers have the option of tuning in to frequencies on which ARI is being broadcast or listening to regular broadcasts. ARIAM is a more sophisticated development of ARI which includes an information collection component. Roadside sensors transmit weather and traffic information to a central processor and this information is used to supplement traffic details from other sources. ARIAM is likely to continue in parallel with the development of the Radio Data System Traffic Message Channel (RDS-TMC)—the direct use of car radios as receivers for digitally encoded traffic information—since it may be used for variable message signs and for people without an RDS-TMC receiver. Technical details of the RDS-TMC system will be discussed in Section 3.

On-board navigation systems: This category of system provides drivers with information about the current location of their vehicle, and how this relates to their destination. Some systems can also provide limited advice on "best routes". The information is calculated by a self-contained unit in the vehicle, which does not require any external link to the roadside infrastructure. These systems can be divided into three main types: directional aids, location displays and self-contained guidance systems. Directional aids typically use dead reckoning from measurements made by distance and heading sensors to continuously compute a vehicle's progress from a known starting location. Location display systems inform the motorist of current position on an in-vehicle display unit, which also shows the actual road network. However, these systems do not offer advice on the best route. Examples of these systems include the USA's Loran-C system, Navstar GPS, and Philip's CARIN, which is illustrated schematically in Figure 1.3. The CARIN system provides the driver with information about the vehicle's current location within a road network (see Thoone, 1987). The vehicle is equipped with direction and distance sensors which allow an in-vehicle computer to calculate the vehicle's position on a map of the road network which is stored in the computer's memory. This information is displayed to the driver on a small electronic map.

Self-contained guidance systems provide the driver with actual routing advice, as well as vehicle location information. Such systems require that a vehicle is equipped with a vehicle unit which contains a comprehensive description of the road network, together with an algorithm to compute an optimum path through the network.

Externally linked route guidance systems: This category includes all electronic route planning and route following aids which have a communications link from in-vehicle guidance equipment to an external system providing network or traffic information. These systems have the potential to take account of "real-time" traffic conditions, and the extent to which they do so depends upon the system under consideration. These systems can be further divided into two main categories. Those linked by a long-range communication or broadcasting channel to the traffic information service, such as those linked to a paging system or radio data system (RDS), and short-range communications systems which are linked to the roadside infrastructure, such as ALI-SCOUT/EURO-SCOUT.

Automatic vehicle identification systems: Vehicles are uniquely identified as the on-board technology activates a permanent logging device in the road without requiring any action by the driver or an observer.

Automatic vehicle location systems: AVL systems are similar to on-board vehicle location systems, except that the emphasis is upon providing vehicle location information to a central control, rather than to the vehicle driver.

RTI TERMINOLOGY AND CONCEPTS

This general introduction to key RTI terminology and concepts requires a cautionary note. As in many actively developing areas of international research there is, as yet, little standardisation of terms. The Glossary can be referred to if clarification of terminology is needed at any point in these descriptions. The Glossary, and in particular the concept maps, are designed to help clarify terms used in this text and also indicate alternative European and American terminology. In addition to the following definitions, there is a general discussion of each term that will help to set it in context.

Road transport informatics

As we have already discussed, RTI is *the application of information technology (IT) to road transport management problems.* Three relatively recent developments in information technology have made RTI possible:

- The miniaturisation of components has been made possible through recent developments in silicon chip technology. This has made in-vehicle processing of information through on-board microprocessors feasible.

- Developments in the field of information transmission systems has simplified and increased the possibilities for vehicle-to-roadside communication. This has largely been through the use of infra-red and microwave transceivers.

● General developments in the theory of information technology, particularly data structures and transmission, have encouraged the development of new applications. The establishment of agreed forms of communication between different systems and the move towards international standards has proved crucial to development.

The physical environment

This is, simply, the *physical context within which an RTI system will operate*. The most significant features of the physical environment are, of course, the road network, and the vehicles that use it (Figure 1.1). The road network may be urban or rural, regional, national or transnational. It may simply consist of one section of toll road, or a large and complex arrangement of city streets. This road network includes the road, roadside, local and central traffic control centres, and can be labelled as the infrastructure. Traffic density and the type of road-user (leisure, business, haulage etc.) will vary between locations and different times of day.

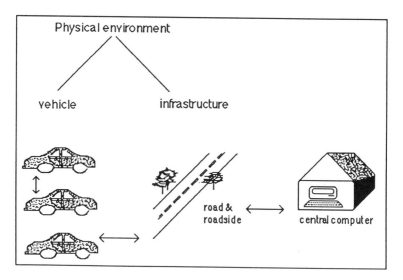

Figure 1.1   The physical environment in RTI

The fact that an RTI system is in operation will, in itself, affect traffic flow. The effects on traffic flow of the use of RTI systems is, of course, a central issue in RTI. Since few systems have been implemented on a large scale, there has been little opportunity for detailed evaluation of their effects. The interaction between system and environment is likely to be complex. For example, the impact of any one traffic management system on the traffic flow will be a complex function of variables such as:

- the number of cars equipped to use the system;

- the extent to which drivers act on the advice given;

- the effectiveness of the way in which driver advice is computed by the system;

- the effectiveness of the way in which driver advice is delivered to the driver.

In describing the effects of a system on traffic flow it is important to draw a distinction between vehicles that are equipped with the technology or are participating in the system, and those that are not. Several levels of participation are possible.

- *Passive* vehicles either gather no information or only do so on an occasional basis through scheduled radio broadcasts.

- *Active* vehicles have a one-way interaction with the system, continuously and selectively gathering relevant information, transmitted to the in-vehicle system.

- *Responsive* vehicles have a two-way interaction with the system. They are equipped both to receive from and to transmit information back to the infrastructure.

In a vehicle guidance and traffic control system such as ALI-SCOUT/ EURO-SCOUT, all cars are likely to either be passive (cars not equipped with the system) or responsive (cars equipped with the system). A further level of differentiation concerns the level of response from motorists to the information received. Does the receipt of information lead to a modification of behaviour? This question of the acceptance of new systems by road users will be discussed in Chapter 2.

RTI applications

An RTI application is *a function within the field of road transport management which might, potentially, be fulfilled using information technology.* Applications have so far been discussed in only very general terms, for example "parking management". However, when an application is being considered in a specific situation it will be necessary to describe it in more detail. This detailed description will comprise, in effect, the technical requirements for any system that might be used to fulfil the application.

RTI systems

An RTI system is *a means by which an RTI application might be fulfilled.* Describing the application specifies to *what* use RTI can be put in a particular locality, whereas describing the system specifies *how* this is to be achieved. Assume, for example, that it has been decided that a parking management application will need to communicate parking information to vehicles within a fifty kilometre radius of the town centre. It is possible to imagine several different systems that will be able to transmit this information. An appropriate system might transmit the information from a single radio transmitter. Another system might rely on a combination of dedicated transmitters at special locations.

System components

An RTI system can usefully be divided into several component parts. These will be either in-vehicle, or part of the system's infrastructure. The infrastructure is defined as any part of the system that is not installed within the vehicle. This may include roadside components, such as the beacons used in the ALI-SCOUT system, and any central processing capacity that the system might have. The components present in any particular system are dependent on the application that the system is designed to fulfil, but all systems are likely to include a need for more human–computer interfacing (HCI), some sort of information processing and/or storage capacity, and one or more information transmission systems. Many systems will also include some form of sensors. We will now discuss each of these in a little more detail.

Human–computer interface

This is the critical component that makes it possible for a person to enter information into, and extract information from, the information system. The most obvious role for the interface is to communicate road traffic information to the driver. However, some form of interface is required at each point at which people must interact with the system. If, for example, the police or the road traffic organisations input information about traffic conditions into the system, a well-designed interface is necessary to make this possible with maximum efficiency and accuracy. Similarly, where drivers need to enter data into the system, for example inputting a destination code into the ALI system, an input device that is simple to operate is required. An example of such a device is the key pad of the ALI system . Interface design is an important consideration in the planning of an RTI system as a whole and we will return to this issue in the next chapter. (Note: Throughout our discussions we shall use the abbreviation HCI to refer to both the product—*the human–computer interface* —and to the process—*human–computer interaction.*)

Processing and storage capacity

Almost all RTI systems require some sort of information processing and storage. Even a relatively simple application such as automatic toll debiting needs information-processing capacity to decode the driver identification messages transmitted by passing cars and to debit the driver's account accordingly. Traffic guidance and management systems such as the ALI-SCOUT/EURO-SCOUT system need considerable computing power to translate the messages from roadside beacons into an overall picture of traffic conditions and then, on the basis of this picture, to produce routing recommendations that can be transmitted to drivers.

As we have already noted, recent developments in computer hardware have made RTI a feasible traffic management option by providing large amounts of processing capacity in compact and affordable units. One development that is particularly useful for systems that require in-vehicle data storage is the CD-ROM (read only memory on a compact disk). CD-ROM is used by the CARIN system to store a road map of a particular locality. The large capacity of the CD-ROM and the speed with which the information can be retrieved from it makes it ideally suited for this task.

Transmission system

This provides a communication link between vehicle and infrastructure, between different components within the infrastructure, and, depending upon the application, inter-vehicle communication. Transmission systems are, therefore, the backbone of any RTI system. There are particular design problems associated with data transmission in a road traffic environment. These include difficulties associated with the transmission between mobile vehicles and a stationary infrastructure, and communication within a built up area. For safety reasons, the quality requirements of data transmission in RTI systems, high accuracy and low bit error rates are likely to be more stringent than in many other fields and on a comparable level to financial institutions.

Due to the importance of information transmission, it will be the focus of the majority of examples used in Sections 2 and 3.

Sensors

A variety of sensors play an important role in many RTI systems. We will define sensors as *any component that extracts information from the physical environment, be it vehicle or infrastructure.* This definition includes a large number of possible devices. For example the CARIN system relies upon in-vehicle direction and distance sensors to provide the necessary information to calculate the vehicle's location. Roadside sensors are able to detect a diversity of aspects of the road traffic environment including the volume of traffic on the road, exhaust emissions, and current weather conditions.

A range of other components and sources of information are available to the RTI system designer. Figure 1.2 gives a general overview of how the components may be organised. Some of these components are illustrated by the CARIN system, illustrated in Figure 1.3 and by the Universal Vehicle Information System (UVIS), illustrated in Figure 1.4.

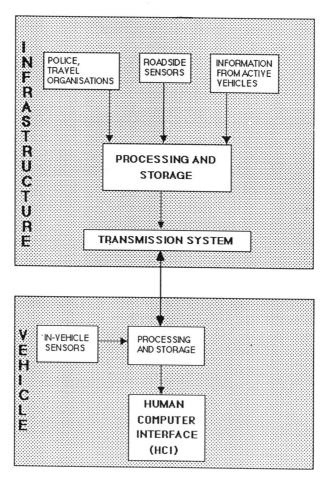

Figure 1.2   Components for a driver information system. Note: Although processing and storage are represented as one system in this figure, and elsewhere, the memory system addressed by the computer can, of course, be physically separate from that computer

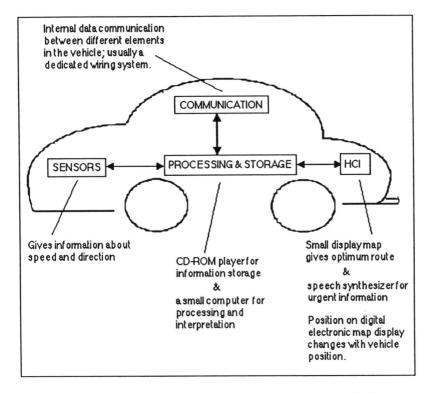

Internal data communication between different elements in the vehicle; usually a dedicated wiring system.

COMMUNICATION

SENSORS

PROCESSING & STORAGE

HCI

Gives information about speed and direction

CD-ROM player for information storage
&
a small computer for processing and interpretation

Small display map gives optimum route
&
speech synthesizer for urgent information

Position on digital electronic map display changes with vehicle position.

Figure 1.3 An in-vehicle driver information system: CARIN

The product

A product is *a physical component that can be used to implement an RTI system*. For example infra-red transmission may be the chosen technique for disseminating parking information. The next decision to make would be whether to use the infra-red transmission technology developed by one company, or the technology developed by another company. These different products have distinct performance characteristics, which along with factors such as cost, need to be taken into consideration when selecting the product.

Although we have drawn a definite theoretical distinction between system and product, in practice the difference between the two can become blurred. Because this text is intended primarily as a conceptual introduction rather than an exhaustive technical manual, the term "system" will now be used loosely to mean both system and product. System or product will only be referred to specifically when the distinction between the two is relevant.

The integration of in-vehicle components is illustrated in the CARIN system, shown schematically in Figure 1.3. This vehicle-based electronic route navigation aid is not a dynamic system, and therefore does not account for

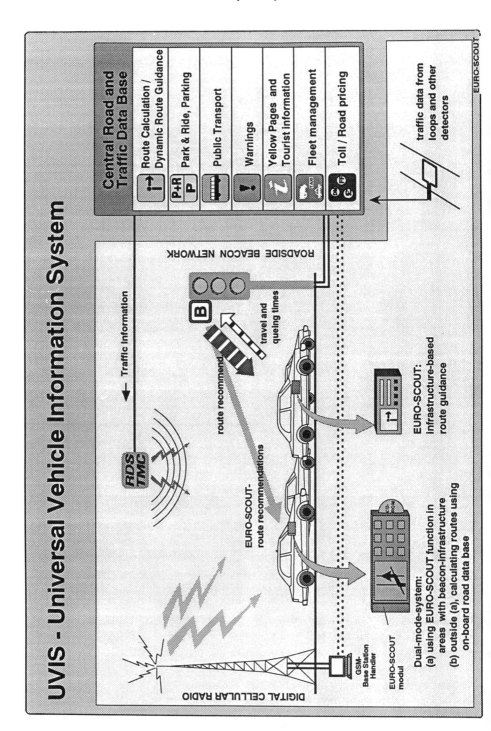

"real-time" traffic conditions, but provides an illustration of the relation-ships between the components of the product. A related idea is the Universal Vehicle Information System (UVIS) from Siemens-Plessey, which uses the EURO-SCOUT route guidance system. Traffic information is col-lected by a central data base from a number of sources, including road loops for the detection of traffic density, in addition to 'static' information about road tolls, parking, public transport, road hazards, and tourist information. This information is available for transmission to vehicles by radio (using the traffic message channel) and by roadside beacon directly to the EURO-SCOUT route guidance module. The in-vehicle EURO-SCOUT unit will also be able to use maps and other information stored on CD-ROM. The flow of information within UVIS is illustrated in Figure 1.4.

SUMMARY AND OVERVIEW

The contents and approach taken here can be illustrated by Figure 1.5, which shows the different aspects of evaluation that have to take place in the con-text of traffic, transport and telematics. It shows how we can relate explicit products to political and economic goals. Such goals would be defined on a macro level (national, continental) as well as on a micro level (organisation, company).

The starting point of almost all systems and regulations that are intro-duced is the goal of obtaining behavioural change. An example here is the behavioural change obtained by providing information for better use of alternative modes of transport as a means of reducing congestion and strain on the environment. Another example is the introduction of highway con-trol systems with speed recommendations as a means to change driving behaviour and improve safety and throughput, as well as improving driving comfort and strain on the environment. The objectives of most RTI applica-tions have to be translated into some kind of analysis of the possible effects upon the behaviour of the individual road user and the behaviour of organisations responsible for traffic management. Chapter 2 describes aspects of behavioural change related to safety objectives and safety related applications.

How users can specify requirements in terms of a general model and interfaces is described in Chapter 3. Users in this context often means repre-sentatives of users of transport and traffic facilities such as politicians, authorities, or the management of companies involved in the development and manufacture of RTI products. These users are involved in defining the

---

Figure 1.4 The Universal Vehicle Information System, which transmits information from a central road and traffic data base to vehicles. (Figure reproduced by courtesy of Siemens-Plessey.)

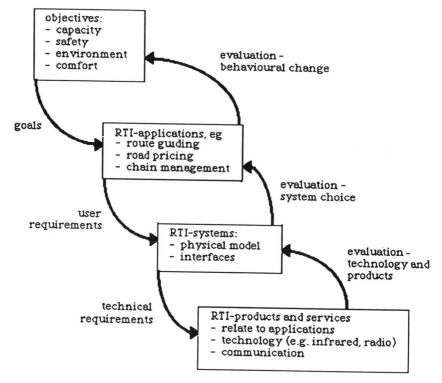

Figure 1.5   An outline of the iterative RTI evaluation process

requirements for applications and for testing these requirements in practice. Users are also involved in the evaluation of the systems that can be defined for applications with respect to the defined requirements.

Suppliers of technology and products are involved in the selection of products and services. We need to give suppliers or developers the option to show the way they can meet the requirements. In practice this means that not all requirements are always met and that different products or technologies meet the technical requirements in different ways. Chapter 4 shows how a number of different technologies meet requirements in quite different ways.

Such a process is not usually started without having some knowledge about the market of RTI suppliers and their available technologies in order not to define highly unfeasible projects. Chapter 6 gives an overview of issues that relate to radio systems that are an important technology for RTI applications.

The whole process described here is highly iterative. This means that requirements have to change often several times due to impossibilities to meet requirements because of technical or financial reasons. At every stage

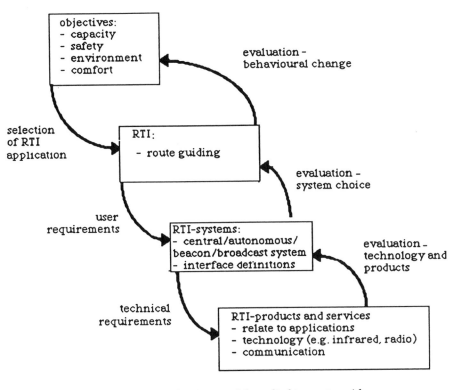

Figure 1.6 The evaluation model applied to route guidance

politics may play a role (standards, existing laws and regulations, financing, etc.).

Costs and benefits are indirectly included in this model. Cost can be determined on the basis of a specific technology or a number of defined and known products and services (often already existing). One should also include the costs for making behavioural change possible (e.g. introduction and implementation costs).

Benefits are more difficult to quantify, as they are related to aspects like environment, comfort, safety. In general one has to limit oneself to the economical gains of improved throughput and savings from costs for transport facilities (roads, gas, etc.). Usually the costs and benefits are determined for individual travellers only on economical grounds. An extensive example of this analysis is given in Chapter 7. We should stress that such types of analysis usually take place several times during the process of RTI system development.

An important RTI application is route guidance. The process described in general terms in Figure 1.5 is applied to this specific application in Figure 1.6. Examples and cases in this book often relate to this innovative application of mobile communications. Chapters 5 and 7 discuss different aspects of route guidance in detail. Figure 1.6 is not complete and a few remarks are appropriate at this point:

- *Route guidance:* The term route guidance is often defined in different ways. For that reason one should always define to what user the term relates: individual drivers or road authorities. In earlier documents on RTI the term also related to vehicles that were automatically guided. This term is now often named driver support, as the role of the driver is still far from superfluous in such systems.

  After definition of the scope one may determine requirements. Requirements may involve national coverage or independence from other infrastructures. Projects which implement route guidance systems should be evaluated in terms of the observed behavioural change in relation to the stated goals.

- *Systems for route guidance*
  It is necessary to define a general physical model including the following aspects:
  central, decentral or different;
  the network aspects;
  interfaces on the basis of functionality;
  coupling of application and network;
  management issues for the complete system (e.g. information management, network management, traffic management, etc.).

  After evaluation with respect to user requirements the physical model can be determined.

- *Products and technology* On the basis of a system choice the different products and services can be implemented. The costs of the application can now be determined and compared to the expected benefits. The technology and the products are evaluated on the basis of their technical performance, and the system evaluated with respect to its applicability. The specific route guidance application is evaluated with respect to its ability to induce behavioural change for individual drivers, for companies (if applicable) and for a region as a whole. Only at this point will we be able to compare the gains made by implemented system against the goals of the original design.

RTI has a wide range of potential applications and a large number of RTI systems are either available or under development. This diversity of applications and systems means that decisions about how to apply RTI technology in a particular situation need to be taken carefully and methodically.

A typical scenario likely to gain considerable benefit from the implementation of an RTI system might be as follows. A large town has serious traffic congestion problems and limited parking availability. It also has a large number of historic buildings which are both a tourist attraction and a hindrance to further development of the road system. Money is available for improving traffic flow in and through the town, but there are good reasons for finding an alternative to simply developing the road network.

There are a variety of possible ways in which RTI can be used to help relieve some of the town's traffic problems. Also, for each application there are likely to be various possible systems either under development or ready for implementation. Deciding how to use RTI to solve traffic problems in a particular road network or locality will involve the consideration of a large number of factors. These range from the technical requirements of specific applications to the legislative and political constraints on implementing a particular system in a particular locality.

This text describes ways of helping to solve traffic problems through the use of road transport informatics, and provides a systematic approach to the evaluation of RTI technology in a particular setting. We have called this the *RTI assessment procedure*, and will often refer to it simply as the "assessment procedure". It starts by offering a technique for deciding which RTI application is likely to be most beneficial, then looks at systems that might potentially fulfil this application. The next stage compares the technical requirements of the application with the technical features of relevant systems. Systems are then assessed for their feasibility of implementation, and are finally analysed for cost effectiveness. Before describing the assessment procedure in detail, we shall complete this introduction by emphasising the need to take account of the behaviour of the operator of the system.

2 ISSUES IN RTI: INTERACTIONS BETWEEN SYSTEMS AND USERS

ISSUES IN RTI DESIGN

It is important for users to be aware of current RTI developments and how they will impinge on their day-to-day lives. This discussion is primarily an introduction to RTI for users and developers rather than for primary designers, although this chapter will point to the importance of evaluating alternative designs as a part of system development. The issues raised here, which have been the focus of recent debate and development, are interrelated and involve one or more of the following factors: people, technology and economics. This part of the discussion will draw heavily from the consideration of these issues by Bonsall and Bell (1987) and Shneiderman (1992).

Human–computer interaction

HCI (human–computer interaction, or perhaps preferably, human–system interaction) involves the management of the interaction between the system and its users and is a central issue in RTI design. It affects both the technicians who operate and maintain the system and the drivers who receive information from it. For example, it is important that an RTI system, once established, is sufficiently transparent in its operation, and in the way it is programmed, to allow on-going maintenance and development. Computer technicians working in RTI, as in other fields, have benefited from recent developments in the human–computer interface. Graphic editing, for example, has reduced the repetitive nature of software maintenance.

The relationship that has received most attention, and deservedly so, is the nature of the interaction between the driver and the system. This is, of course, a complex area but can be divided into three sub-issues:

- the depth of information supplied to the driver;

- the extent to which the system affects the autonomy of the driver; and
- the user-friendliness of the interface.

A thriving industry is currently working on the principles of the optimal human–computer interface, and the design of the interface between RTI system and the driver should be viewed as a part of this endeavour. An interface that delivers information efficiently to the driver is not only more effective, but is also more likely to be used. A route guidance system, for example, may not be used at all by the driver who finds the controls difficult to manipulate, the information in the display difficult to see, or the advice difficult to understand. General principles of the design of interfaces can be found in guides such as Helander (1988), which contains particularly relevant chapters on the design of screen layouts (Tullis), the quality of the image displayed on screens (Snyder), the use of operator input devices (Greenstein and Arnaut), and, of increasing importance, the use of speech synthesisers for message delivery (Streeter). To take account of the behaviour of the end-user when designing systems in general, Gould (1988) advocates the application of four principles, derived from work on standard computer screen displays, but just as applicable to the design of in-vehicle interfaces. These principles are:

- observe the system operator, through interviews, surveys, participative design, etc., in order to understand the task as seen from the operator's perspective;
- ensure that all aspects of the design process evolve in parallel;
- test the prototype systems early and continuously during development, and observe the user's qualitative and quantitative performance; and
- use an iterative design process to modify the evolving system on the basis of the result of user performance tests.

A detailed description of the process of user-centred design can also be found in Shneiderman (1992), which includes the basis of a user-evaluation questionnaire written initially for assessing the usability of standard interactive computer systems, but which could easily be modified as an assessment tool in RTI system design. User-centred design should not be regarded as an optional extra by system designers. It helps engineers design effective products. One reason for the success of these principles is that designers work more carefully and with a greater overall view of the use of their product when they know that a usability test is part of the design process.

It is a straightforward matter to evaluate the human factors associated with operation of any complex system. Schneiderman's guide describes the five measures central to this evaluation, and suggests systems for which they will be most pertinent. An important part of the evaluation is to set up benchmark tests which are characteristic of the intended performance of the system, which involve all of the operating controls of the system, and which are held constant during the testing by different users. The five measures are:

- *Learning time:* How long does it take for typical users to learn the commands necessary for operation of the system?

- *Speed of performance:* How long does it take for the typical user to perform a set of benchmark tasks?

- *Error rates:* How many errors do users make in performing these benchmark tasks, and what kind of errors are made?

- *Stability of learning:* How well do users remember what they have learned?

- *Subjective satisfaction:* How well did the users enjoy using the system?

The development of any system intended for operation by the driver of a vehicle requires the use of benchmark tasks which take account of the special circumstance of driving, of course. Driving requires "multi-task performance" in that the operator is often performing several tasks at once, with a hierarchy of importance associated with those tasks. Control of the vehicle itself requires the performance of several component tasks (vehicle controls involved with direction and speed) and at a higher level the driver needs to monitor the behaviour of adjacent traffic, and plan the route to the intended destination. Adding a further task to non-RTI driving could prove critical to performance of any of the driver's existing tasks, if it requires attention to directed away from the primary task of vehicle control. The evaluation of the human factors associated with the operation of RTI systems should therefore make use of benchmark tasks performed during driving or simulated driving, in order to determine any interactions between tasks. The requirements of such a benchmark task for assessing the load placed on the driver by the introduction of an RTI system have been described by Parkes (1991). For example, a task which asks about the effects of providing route information might compare a number of methods of providing information over a fixed route using a range of performance measures. In delivering route information through an in-vehicle screen we are necessarily distracting the driver from the primary task of dynamic vehicle control, and any system must be designed to minimise this distraction while providing useful information. Parkes estimated that such a route information system can take as much as 22% of visual scanning time from the scene ahead, at least in drivers relatively unfamiliar with the system. In addition to this potentially dangerous loss of attention there may also be an increase in stress associated with the gathering and use of information from these in-vehicle systems.

The development of systems for specific uses will place different emphases on different measures. For example life-critical systems, such as those used in nuclear power stations, in air traffic control, or in monitoring medical situations, should be error-free. The liability of the user to errors would be an essential part of the evaluation of such systems, and long training periods would be acceptable in order to eliminate operation errors. In contrast, some commercial and industrial systems accept higher error-rates but require rapid learnability. It is expensive to train the operators, and errors in sales and banking systems are regarded as being acceptable. The development of RTI systems with human–computer interfaces will also require emphasis on different performance measures according to the specific application area. Errors will occasionally be life-critical, as with emergency hazard systems, for instance, but with route guidance and parking information the consequence of error will be an increase in operator frustration. This in itself may work against the intended implementation plan. If a driver is given false information by a route guidance system, not only will he experience an increase in driving time but that particular vehicle will occupy the road for a longer period than necessary, and if the error is later detected by the driver his/her subjective valuation of the system will decrease. Future route advice may then be ignored.

There is a need to find a balance between prescribing the driver's actions, providing an explanation for these prescriptions, and providing the driver with general information about road traffic conditions. The argument for providing drivers with a detailed account of the reasoning behind the advice given by the system is that this will increase both the drivers' feelings of control over the system and the probability of their accepting and acting on the advice given. However, there is a limit to the amount of information that drivers can absorb whilst driving. Even when giving total attention to an incoming message we have a natural limit to the amount of information that can be received and acted upon. When the primary focus of our attention is the task of vehicle control, then this limit is severely reduced.

Simple messages are more likely to be effective than more detailed and more informative messages. In some applications, giving detailed information might even lead drivers to act inappropriately. This is illustrated, for example, by the level of detail in warning signs. The hazard warning "Danger Ahead!" is more likely to result in drivers reducing their speed than a more specific warning which may require more processing time on the part of drivers or which may be simply rejected by drivers as unimportant. Similarly it has been suggested that when motorway traffic is guided by variable message signs, the driver should be set a maximum speed limit with no justification offered. The reasoning behind this suggestion is that if drivers are informed about the conditions ahead, they may select divergent speeds which may in turn result in a more dangerous situation.

The effectiveness of any system will be, in part at least, a function of the acceptance of the system by prospective users, particularly if alternatives are available. In the case of certain RTI systems their use may be completely optional, as with route or parking information. If the driver can choose to not use this information then we must ensure that it is perceived as being useful. Bonsall *et al.* (1991) have investigated drivers' reactions to route guidance systems with a questionnaire survey. As well as asking drivers whether they were likely to use guidance systems, and the systems and circumstances in which they would be used, drivers in Berlin were asked about their actual use of the LISB system. LISB is based upon ALI-SCOUT, and has been on trial for some time. The acceptance of route information is dependent upon the perceived quality of the information, with drivers being intolerant of advice that contradicts their own knowledge of the route. A fifth of the drivers questioned changed their normal route to and from work in response to advice from the LISB system, but an equal number declared that they would not vary their route even when advised to do so.

The extent to which it is acceptable for the system to impinge on the autonomy of the driver is the focus of considerable debate. As yet automatic vehicle piloting has not reached a stage of development where it can be used as a replacement for a human driver. Even if it had, however, past experience has shown that many drivers would reject such a system because they would be unwilling to tolerate the loss of control. There may be cultural differences here. Europe has not taken to the small control loss encountered with an automatic gear shift mechanism, yet these systems are widespread throughout North America. A useful compromise is for the RTI system to take over the routine vehicle control, and provide additional information to the driver, who can then decide whether or not to act upon it. This is essentially the way that aircraft piloting has developed. There is a precedent for this increased machine control in the use of automatic gears. Society's tolerance to such systems may vary between different cultures. Vehicles with automatic gears are the norm in the United States, but occupy a lower proportion of the market in Europe. The nature of the application also affects acceptance levels. Applications such as automatic debiting are likely to prove more acceptable. It is unlikely that drivers will want to feel in control of such a system's operation, and, if the driver were to be given control it would significantly reduce the system's usefulness.

Making the interface between the system and the driver easy to use is essential both from a safety point of view and to ensure that drivers make full use of the system. It is also important to make the driver feel comfortable with the interface; "techno-fear" (the fear of technology) is a genuine barrier to the addition of new technology to the familiar vehicle controls. Generating this feeling of comfortable use is most likely to be achieved when the system is designed around the natural abilities and limitations of the intended user. As well as the obvious limitations to our use of effector systems—the number of hands and feet we have available, for example—there are hidden limitations to our psychological abilities to receive and use information. A display system will not be used optimally, for instance, if it demands that the driver act upon briefly shown information while maintaining attention on vehicle control and changing traffic circumstances. A display must be available until the primary demands upon the driver allow a redirection of gaze and inspection of the incoming message. We also have memory limitations, and if information is to be acted upon systems must either present information which can be apprehended in a short inspection and then remembered, or support the driver's memory by retaining the information on the screen or on a recallable screen display. Factors such as limits to the driver's abilities to gather and remember information will be taken into account if the system is to be declared as user-friendly.

Recent technological developments have endeavoured to lower techno-fear by simplifying input and output devices to the system. Devices which allow the user to employ standard human communication techniques such as pointing or speaking increase the user friendliness of an interface. Such devices include touch sensitive screens, computer voice production and voice recognition systems. Voice recognition is potentially a very interesting input mechanism for RTI but currently only a limited range of commands can be recognised by machine. The difficulties of natural language comprehension by computers are well documented elsewhere (e.g., Briscoe, 1987; Reichman, 1985). The obvious benefits and wide area of application of a voice controlled system are proving a potent research stimulus but in the foreseeable future the "optimal solution" is likely to incorporate a combination of several different input modes.

Institutional issues

Numerous issues arise within an institution as a result of the implementation of information technology. An increased use of technology often results in the redefinition of job descriptions for individuals and so the reaction of personnel to these changes can be negative. A negative reaction is often based upon the real or imagined prospect of redundancy, retraining programmes and skills redefinition. However, such retraining is likely to become an ever-increasing reality in the future. Therefore employers and those responsible for implementing changes are looking more and more to create an atmosphere in which retraining in information technology is perceived in a positive manner. It is expected that future generations will be far more flexible with regard to information technology as classroom computers become a regular part of children's educational experiences (see, for example, Underwood and Underwood, 1990).

Another issue stems from the fact that a large number of organisations are involved when RTI is being implemented: not only vehicle manufacturers, electronics industries, PTTs and private telecom providers, but also the public, the local, regional and national governments (and in the case of Europe, the EC), the road authorities, the police, etc. In this complex setting some organisations have to change their roles and responsibilities. As an example, the provision of traffic information becomes a combined responsibility of police, road authorities and private organisations. So these organisations have to provide a new service with new technologies in a new organisational setting. It is especially these new roles and responsibilities, as well as these unfamiliar ways of collaborating, that are presently giving rise to much debate.

Locating processing capacity

Some form of information processing unit is likely to be present in all RTI systems. When traffic management systems such as ALI-SCOUT/EURO-SCOUT (described in the previous chapter) are fully implemented, they are designed to receive information at regular intervals from a large number of vehicles, and on the basis of this, compute route finding instructions that are specific for each driver. This requires large stores and rapid processing capacity and also puts considerable load on the communication channels between the central processor and the vehicles.

In order to overcome these problems of data processing, storage and transmission there is an increasing trend towards decentralising the point of processing in the system (Figure 2.1). This is achieved by locating processing capacity within vehicles and/or at lower levels of the infrastructure. In large systems, it is possible to partially decentralise processing from a single

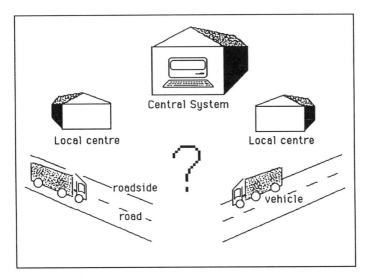

Figure 2.1   Where should the control of information processing be located?

control centre to several, local "outstation monitoring units". At a vehicle-by-vehicle level, this decentralisation makes it possible for each vehicle equipped with the system to receive an identical message, and then to extract from that general signal information that is relevant to the individual driver on the basis of the intended destination.

The relocation of at least part of the processing capacity of the system will significantly decrease the size and complexity of the load on the communication channels and reduce the transmission problem. The disadvantage of decentralised processing is that it makes fault detection, correction, and system modification a more complex task. However, this is often a less serious problem than that of transmission if remote reprogramming is possible.

Optimisation

This issue entails choosing the most appropriate method for solving complex problems, like finding optimal routes or scheduling parking spaces, and also proving the operational integrity of co-ordinated complex traffic signal layouts. For example, in the development and choice of RTI systems, several concepts exist in the design and modification and these are not necessary mutually exclusive:

- *Satisficing behaviour:* This is an economics concept often linked to optimisation. While a high-order goal may be to reach an "ideal" or "optimal" solution to any one problem such solutions are costly in time and money. In practice suboptimal but feasible solutions are often accepted. Such solutions are seen as "satisfactory". This satisficing behaviour predominates when the problem is very complex, with partial, satisfactory solutions being adopted in an *ad hoc* fashion in response to a particular bottleneck.

- *Fragmentation:* Complex designs which are easily compartmentalised can be optimised by fragmenting tasks and altering the relevant parts.

- *Heuristics:* Unlike the above two strategies, there exist global heuristics; rules of thumb which offer a solution to the whole problem/design. These simple rules do not necessarily produce the optimal solution, but for certain systems or application areas, they have proved to be adequate, simple and cost effective.

- *Combinatorial methods* (or integer programming methods): These methods may well replace the use of heuristics, as a result of the progressive increase in computing power and the falling costs of powerful workstations.

Interaction with demand

What is the relationship between increasing information technology (IT) and transport ? Does IT stimulate or replace the use of physical travel ? IT developments, such as in-vehicle route guidance systems, could stimulate travel by reducing congestion and thereby attracting road use. Alternatively other IT developments such as "armchair shopping" or tele-conferencing may reduce the need for travel. We should therefore make a clear distinction between the use of IT in relation to present ways of obtaining mobility (e.g. traffic management), and the use of IT to provide an alternative to current practice (e.g. tele-shopping).

Standardisation

There is a recognised need for standardisation of both information and technology. The Open Systems Interconnection (OSI) model, which is described in Section 2, illustrates this approach. On a European level this is necessary for optimum progress with minimal redundancy (quite apart from the advantages inherent in increased international communication). However, questions have been raised as to whether the costs relating to the organisation of agreements on standardisation are realistic in comparison to the complete market for the RTI products. It may well turn out that it is much cheaper to use patented products, as the size of the market does not allow for an open approach, and to accept that the market is governed by *de facto* standards.

Financial cost

In order to calculate initial *outlay of capital*, the system may conveniently be broken down into the following components:

- in-car processor and storage
- in-car sensors
- driver–machine interface
- traffic or transport central processors
- IT for traffic and transport management organisations
- road infrastructure costs, e.g. beacons, sensors.

In addition to initial outlay, it is necessary to take into account the *on-going costs* of communication within the infrastructure. The norm is to use TA (Telecommunications Administrations) facilities for this. Important issues are also those that are related to maintaining the system. The estimates of revenues depend very much on the market size and approach as well as on the growth potential of the services. Several alternative economic scenarios for RTI systems have been discussed by Marco and Flowerdew (1991).

ESTIMATING BEHAVIOURAL ADAPTATIONS

A priority of road safety authorities is to assess the safety programmes and countermeasures developed. One of the three primary objectives of the introduction of RTI is to improve road safety, recognising the huge human and financial costs associated with road accidents. Accordingly, a large number of projects have investigated the use of RTI systems to enhance the safety of road users. As innovations of any kind are introduced into the road transport system, however, the behaviour of road users will change. If the system is perceived as being safe, then driving behaviour becomes more risky, and the benefits of an innovatory safety measure will not be optimal. As an introduction to the problem of how to understand the actual effects of an engineering innovation upon the behaviour of the users, the present discussion outlines a model of behavioural adaptation which we have developed elsewhere (Jiang *et al.* 1992; Underwood *et al.* 1993).

The aim of any RTI implementation is to enhance our use of the road network and is dependent upon behavioural change. One of the clearest forms of this change results in an improvement of the objective safety of the roads, but other general goals are also dependent upon behavioural change. If we want a route guidance system to work optimally, then drivers will need to change the way that they think about their route decisions. The study of responses to such a system, by Bonsall *et al.* (1991), suggests some sources of resistance to behavioural change that must be taken into account upon general implementation of the system.

Unless we have the basis for predicting the behavioural adaptations of road users, following the introduction of road-traffic safety programmes and engineering innovations, the effects of new safety measures will not be known. At present we simply do not have such knowledge, and the development of an understanding of behavioural adaptation and its effects on road safety is needed for both road safety practitioners and researchers. A report of the Organisation for Economic Co-operation and Development (OECD, 1990) points out :

> " . . . there has generally been a lack of effort to develop an understanding about the process of behavioural change which accompanies road safety programmes. The behavioural change process is complicated, and the reliance on

outcome measures such as accident fatality rates has often precluded the inclusion of data collection which would permit the study of the change process. With a more complete understanding of the process of behavioural adaptation it may be possible to predict how safety programmes will influence the behaviour of road users." (p.117)

We need to not only explain behaviour but must also allow for its prediction as engineering innovations are introduced. The following recommendation has been provided by the OECD Report as a concrete suggestion for road safety administrators, programme planners, and researchers:

"There is a need for further development of theories of road user behaviour to assist in the understanding and prediction of behavioural adaptation. At the same time there is a need for those conducting evaluations of road safety programmes to incorporate theory testing in their research to provide a basis for the development of theory." (p.118)

Such a model should result in improvements in road safety programmes. Our purpose here is to develop a theoretical model for the mechanism of behavioural adaptations, which can describe the process of the behaviour, explain the nature of phenomena, and predict where and when behavioural adaptation is likely to take place.

Definition and assumptions

The definition and assumptions of behavioural adaptation presented below are given in the OECD Report and developed in the context of road safety. They may be broadened to encompass other activities.

"Behavioural adaptations are those behaviours which may occur following the introduction of changes to the road-vehicle-user system and which were not intended by the initiators of the change;"

and

"Behavioural adaptation occurs as road users respond to changes in the road transport system such that their personal needs are achieved as a result. They create a continuum of effects ranging from a positive increase in safety to a decrease in safety." (p. 23)

As well as referring to adaptations which were *not* intended by the initiators of the change, the first part of the definition should also refer to those adaptations which *were* intended. The reasons for adaptation may, furthermore, be opaque, and neither predicted nor understood by the initiators. Our understanding of adaptation becomes even more difficult if some road users adapt as expected while others do not. Such individual differences in adaptation will have serious implications for safety and mobility, and are in clear need of investigation—who is prone to adaptation and who is not, and under what circumstances do safety measures have their intended effects? The basis of a successful engineering innovation is the accurate prediction of the behaviour of the prospective users, and it must be remembered that there will be behavioural adaptations to any innovation, and not just to those which are intended to improve safety.

For behavioural adaptation to occur, it must be assumed that there is feedback to road users, that they can perceive the feedback (but not necessarily consciously), that road users have the ability to change their behaviour, and that they have the motivation to change their behaviour. Behaviour may be changed by first changing the intentions of the drivers, through education about objective risk, for example, through publicity campaigns. Alternatively, it may be changed through implicit learning over a period of exposure to road situations which include events associated with varying levels of risk. Feedback refers to knowledge and information received from the system which results from changes in the road users' behaviour. Feedback, in this sense, is a major component of driver behaviour models. Therefore, our theoretical model for behavioural adaptation needs to include feedback loops, and to take account of motivational factors involving the driver, in order to provide an explanation of changes in behaviour.

A brief review of theoretical models of adaptation

A number of attempts have been made to account for driving behaviour which varies as the road-vehicle system changes, and which can put the driver at increased personal risk. OECD (1990) contains a review of the following theoretical models which attempt to explain behavioural adaptation. The models are mentioned here as an introduction to approaches to adaptations that have been formulated and have an emphasis on the explanation of risk-taking behaviour.

Klebelsberg's (1977) model provides the basis for distinguishing between objective and subjective safety. Objective safety is defined in terms of the probability of loss of an accident under specific physical conditions (e.g. a specified vehicle speed on a road surface of specified friction), while subjective safety is the feeling of safety experienced by the driver. Klebelsberg argues that objective and subjective safety interact, in that a change in one can result in a change in the other, and that drivers aim to maintain an objective safety which is as least as great as their subjective safety. The maintenance of this relationship constitutes traffic-adapted behaviour. The model shows that safety increases on the behavioural level when objective safety increases, without subjective safety necessarily increasing to the same extent. For example, improving the visibility at a high-risk road feature such as an intersection can actually lead to higher road speeds at the intersection. Higher objective safety can be said to result in a higher subjective safety and subsequent higher accident rates.

Subjective risk also plays an important role in Näätänen and Summala's (1974) theory of risk behaviour, which can be regarded as a comprehensive cognitive structure-model of driver behaviour. This model describes the cognitive processes by means of several constructs which are controlled by the action caused by stimulation in a situation, with drivers on the one hand satisfying their motives for reduced journey times by fast driving and on the other hand adjusting their driving to reduce their subjective risk and their encounters with unpleasant traffic situations.

Cognitive theories of risk learning (e.g. Fuller, 1988) have also taken account of the consequences of unpleasant traffic encounters in the long-term modification of driving behaviour. The phenomenon of adaptation is interpreted in cognitive theory in terms of learning, and specifically in terms of avoidance conditioning and threat avoidance. It is suggested that the driver's previous conditioning history plays an important part in determining the level of accident exposure to be tolerated.

Wilde (1982,1988) asks why drivers are prepared to accept a certain measure of objective risk, and the answer is provided in his controversial theory of the mechanism of risk homeostasis. This suggests that the estimated and the accepted risks retain their equilibrium through the mechanism of risk homeostasis which maintains drivers' levels of risk taking. An objectively safe system would attract more risky behaviour than would an objectively dangerous system, as the users modify their behaviour to maintain a constant level of subjective risk. Increases or decreases in safety are only present during the stage of disequilibrium. The model uses the principle of preservation of the accident rate as follows: "the number of accidents in a certain country depends solely on the accident rate which the population is prepared to tolerate and not on the measures taken in the other areas of this control system, at least not over a longer period of time." It expects the estimated and the accepted risks to retain their equilibrium through risk homeostasis. This model has attracted controversy from a number of different directions. A basic question is: to what extent can theories of equilibrium be used as an adequate basis for the problem? In addition, after a theoretical analysis of the theory of risk homeostasis, it was concluded that "there is no reason why exact homeostasis should be expected to occur as a response of road users to safety hardware improvements: partial compensation seems to be the rule instead" (Janssen and Tenkink, 1988). The debate over the risk compensation model is, so far, inconclusive.

Janssen and Tenkink's (1988) own concern with cost–benefit assessments among road users is expressed in their utility model. Reasons for making a trip are found in the sphere of utility maximisation, in which accident risk is considered to be only one of the relevant components. They describe risk taking as a consequence of utility maximisation. Presumably the road user must find a balance between components such as accident risk and driving speed to achieve a total cost for the trip that is minimal. Thus, the risk level obtaining after a change will depend on the structure of the total utility function. It is a product of a utility-maximising process rather than an independent control variable.

The hierarchical risk model of van der Molen and Bötticher (1988) is structured in terms of a strategic, tactical and operational task level. The model was formulated partly as an alternative to the models of Näätänen and Summala, Wilde, and Fuller which are mentioned above. It is suggested that drivers adapt actions on different strategic levels, and on the basis of the environment and psychological processes. Drivers are described as making decisions about route planning, travelling speed and general journey plans at the highest level, the strategical level, and decisions made at this level will influence activities at the lower levels. At the tactical level finer-grained manoeuvres, such as overtaking, will be planned. The manoeuvres themselves are executed at the operational level, where manipulation of the specific vehicle controls is effected—steering movements, gear changes, and other micro-level skilled operations. The attractive features of the van der Molen and Bötticher model are that it provides an overall view of the hierarchy of decisions made by a driver and, secondly, that it allows the generation of specific predictions about alternative behavioural outcomes in terms of subjective probabilities of traffic events and the utilities of alternative decisions.

O'Neill (1977) bases his decision-theory model of danger compensation on the assumption that the driver cannot be considered constant with regard to his driving behaviour, and that he adapts his driving behaviour to changes in the surroundings as a tendency to danger compensation. This "compensation" is derived from the invariable aims of the driver who—assuming he acts "rationally"—tries to maximise the benefit of the action. During the decision processes necessary for this, the wish to avoid accidents also appears. O'Neill assumes that the driver can correctly assess the situation and his own action. From this it is concluded that, as far as the reaction of rationally oriented drivers is concerned, safety measures can do harm instead of help and that danger compensation must be taken into account before safety measures are introduced. However, it is not certain as to how "rational" should be understood in this context, and how far the driver can estimated as a rationally acting entity. The assumption that the driver can adequately judge the accident probability of each of his actions is also uncertain. In addition, there is no description of an individual process as reaction to the introduction of safety measures.

Evans' (1985) model provides a formalism for the influence of feedback upon human behaviour. The general term feedback is chosen in preference to such terms as danger compensation because feedback can, in principle, be of either sign (i.e. have a positive or negative influence) and be of any magnitude. It is also proposed that the actual safety benefit is not necessarily identical to the engineering safety change. As an example of the occasional mismatch between engineering intentions and behavioural outcomes, consider the effects of the following change to traffic signal sequences. Hakkert and Mahalel (1978) examined the effects of the blinking green phase of traffic signals on accident rates in Israel. The green light was set to blink for the last two or three seconds of the green phase to warn drivers of the impending yellow phase. It was found that the installation of the blinking green phase led to an *increase* rather than the intended *decrease* in accidents. Drivers were presumably attempting to use the extra information provided by the blinking green to increase their chances of clearing the intersection before the onset of the red phase, rather than to increase their chances of a safe pass as intended. In this case the engineering safety change had a negative effect in terms of the safety benefit.

The review of theoretical models presented in OECD (1990) leads to the conclusion: "The theories discussed in this chapter are interesting, but unfortunately are at times overly general, or only indirectly related to the concept of behavioural adaptation" and "the definitions are not always clear." The "theories, however, usually do not explain why and which cognitions lead to the expected compensation of objective risk during its change. There is also the problem of how—driving at an unconscious level (i.e., automatic production of skilled control movements)—the unknown objective risk can be adequately assessed and exactly compensated in the case of a change. If it can be assumed that the process takes place at a conscious level, then it is not clear why the driver acts in such an irresponsible fashion and rejects additional safety" (p. 98). Our model is to be developed to address these problems.

Framework of the Jiang *et al.* model

The mechanisms of behavioural adaptation can be seen in Figure 2.2 as elements of various theories in safety research that are used for predicting behaviour. To understand behavioural adaptation there is a need to explore systematic factors that affect the process of behaviour change. These factors can be grouped into a comprehensive model, and the model can be established upon a frame around which we can organise the evidence concerning behavioural adaptation. The framework is presented in the form of a flow chart, which provides a basis for understanding how the concept of behavioural adaptation is incorporated into the theoretical model of road-user behaviour and what the explanations are for behavioural adaptation. In addition, it will provide practitioners with the guiding principles they need to predict behavioural responses to engineering innovation.

The model will explain how the cognition of objective and subjective risk leads to adaptation (see Figure 2.2, blocks 2–4). Decision theory is used to formulate the multi-level decision structure and utility function of users

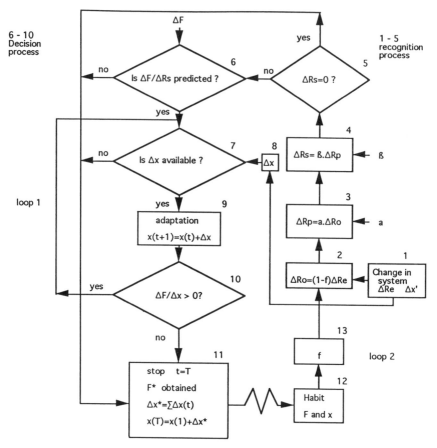

Figure 2.2 A model of behavioural adaptation to change, after Jiang, Underwood and Howarth (1992). Reproduced by permission of Taylor and Francis Ltd from Transport Reviews.

Blocks and variables used in the model depicted in Figure 2.2

Blocks	Variables
1 Changes in transport system, input of model	ΔRe, Expected risk change
2 Relationship between objective & expected risk	$\Delta x'$, Expected behavioural change
3 Relationship between perceived & objective risk	ΔRo, Objective risk change
4 Relationship between subjective & perceived risk	f, Behavioural feedback planner
5 Judgement of subjective risk	ΔRp, Perceived risk change
6 Judgement of risk effects on users' motives	a, Perception parameter
7 Judgement of possible behavioural changes	ΔRs, Subjective risk change
8 Changeable behaviours	β, Attitude parameter
9 Behavioural adaptations	ΔF, Change of users' motivations
10 Judgement of benefit of behavioural changes	Δx, Change of users' behaviour
11 End of adaptation process	$x(t)$, users' behaviour at time t
12 Users' habit behaviours and motivations	F^*, Best satisfaction of users' needs
13 Behavioural feedback to transport system	Δx^*, Total change of behaviour

based on their motives and attitudes towards risk (blocks 6–11). Motivation theory will be used to study motivational factors and explain why and when road users change their behaviour in response to changes in the system (block 10). Control theory is used to describe the whole adaptation process which develops through feedback as a function of time, aiming at maximising the users' utility function (feedback loop 1–2). The behavioural explanation given by the model applies at individual level. It explains why and which factors lead to the expected compensation of objective risk during its change, and it shows that the interaction of users' motivation, expectation and judgement are all relevant in connection with the adaptation process. First, we will describe the cognitive processes involved in behavioural adaptation.

Cognition of changes of system and changes of risk

For behavioural adaptation to occur, it must be assumed that road users can perceive the changes in the system. These perceptions are represented by blocks 1–3 in Figure 2.2. The change to the transport system is presented in block 1, in which ΔRe denotes expected risk change due to system change, Dx' denotes expected behaviour change, and ΔRo change of objective risk. The relationship between ΔRe and ΔRo is indicated by a feedback parameter denoted as f, which will be discussed in detail later.

Behavioural adaptation can be interpreted in the light of the controversy in psychology between cognition of the objective and of the subjective risk (Howarth 1987,1988). Objective risk (Ro) can be defined as the product of occurrence probability and seriousness of an event. Subjective risk (Rs) can be defined psychologically as the expectation of a dangerous event with an estimated seriousness and occurrence probability. Such expectations arise when a goal must be reached without the certainty that it can be safely reached. The more certain a driver feels, the less is his subjective risk. The notion of risk compensation, and its relationship with the effects of safety measures introduced into the road transport system, is discussed in more detail in Underwood *et al.* (1993).

The perception of the physical environment is at each moment influenced by the internal representation of similar situations. It also contains the knowledge of one's own limitations and abilities and interactions with the environment. The internal representation determines judgemental processes by way of expectation, and the attitude of the driver determines the subjective evaluation of the possible results of the behavioural adaptations.

The adaptation model proposes that perception, the process by which road users find out about changes in the system, is primarily responsible for the behaviour change which occurs following an initial response to the system. We present the following formalization to represent the relationship between perceived risk (Rs) and actual objective risk (Ro):

$$Rp = a \ Ro$$

where a is the perception parameter that characterises the possibility of risk perception. Let ΔRp denote the change of perceived risk, ΔRo denotes the change of objective risk, and then we have,

$$\Delta Rp = a \ \Delta Ro$$

This means that awareness of risk depends on the possibility of risk perception, which in turn depends upon the salience of the environmental change (see block 3 in Figure 2.2). If the users have no ability to perceive the objective risk, then $a = 0$; and if users can perceive the entire risk, then $a = 1$. Perception of risk results from observing the road system over time, and detecting changes in other drivers' behaviour and the occurrence of incidents in the road system such as accidents and traffic conflicts.

In the case of some innovations, of course the change is not clearly perceived by the drivers, and so little or no risk compensation is observed. For example, the speeds of Swedish motorway drivers were measured according to whether they were using studded or unstudded tyres (Rumar *et al.* 1976). Cars with studded tyres were driven faster than those with conventional tyres, but the margin of safety was still greater for those with the studded tyres. The drivers were not fully aware of the safety effects of the studded tyres, and did not maintain their objective risk by driving at the maximum speed afforded by their increased safety. They were satisfied with a relatively lower objective risk level. Their subjective risk may even have been greater, of course, as they were travelling faster than other road users without studs.

As the road user's perceived objective risk increases, this will trigger expectancies, in which their attitudes towards risk and their estimation of their ability to master the risk will play an important role. The following formalization represents the relationship between the users' subjective risk and perceived risk:

$$Rs = \beta \ Rp$$

where β is the attitude parameter that characterises the users' estimation of their ability to master risk situations (see block 4). If users have no confidence about their ability, then β may have a value equal to 1 or even greater than 1; and if they have total confidence in their ability to get rid of risk, β may has a value close to zero. Let ΔRs denote the change of subjective risk, and then we have,

$$\Delta Rs = \beta \ \Delta Rp$$

In case of $\Delta Ro=0$ (no change in objective risk) or $a=0$ (no ability to perceive a risk) or $\beta=0$ (total confidence in the ability to handle the risk), then we will have ΔRs=0. That is, there is no change in the users' subjective risk, if there is no change in objective risk, or users can not perceive the change of objective risk, or after perceiving the change they believe that they can cope with it without any actual increase of risk taking. At the end of the process, a judgement of whether ΔRs = 0 is made (see block 5). If it is true, there will be no occurrence of behavioural adaptation responding to the change of objective risk; otherwise, further decisions will be made as to what kind of behavioural change, if needed, is conducted.

The Jiang *et al.* (1992) model suggests that we can account for the causes of behavioural adaptation with the following factors:

- defects in the perception of objective risks, (block 3, parameter a);

- over-estimation of one's own abilities in the mastering of situations with objective risk (block 4, parameter β);

- conscious decision to take an objective risk.

The model suggests that adaptation can be interpreted as a phenomenon which occurs when the driver decides to change an existing action pattern due to a new estimate of subjective risk which results from a new perception or a change in the attitude towards objective risk.

Motivations underlying behavioural adaptation

The driver's motivations are important elements in behavioural adaptation, and will determine the subjective importance of the possible results of behavioural change. The model must take account of the crucial problem of identifying the driver's goals for behaviour at a number of strategic levels, and take account of the driver's understanding of the benefits that will occur as a result of different actions. For adaptation to occur, it must be assumed that road users have the motivation to change their behaviour, that is, there must be a goal set for travel, and the goal must be based on a certain performance tendency. For behaviour adaptation towards a change in risk, it is also needed to assume that the attainment of a goal requires that an objective risk be taken, and that the users must also perceive some level of subjective risk. All behaviour is determined to a high degree by habit and this is certainly true of an over-practised skill such as driving. Reaching a certain destination by means of the chosen vehicle and roads can be referred to as the main motive, but other motivating components are present as well. There is a tendency to avoid objective risk, in which the limits of safety and the avoidance of fear play an important role (Fuller, 1984, 1988). This conflict between performance and safety tendencies is of special interest to traffic psychology.

The driver's motives also play a role in Näätänen and Summala's (1974) "zero risk" theory of driving behaviour. A distinction can be made between excitatory and inhibitory determinants of driver behaviour . Among the excitatory motives are the goals set for the trip, the emotions elicited in traffic situations (including driving pleasure and risk as a reinforcer), and any tendency towards showing off (a public display of one's abilities) or to prove oneself to a peer group. The most important of the inhibitory motives is the subjective risk, and accidents are said to occur when the driver's subjective risk is inaccurately small.

The model accounts for the adaptation by the invariable aim of the driver who tries to maximise the benefit of the action. The control variable is driver's behaviour which is denoted as x in the block, and behavioural adaptation is denoted as Δx. The driver's goal is represented by the utility function denoted as F, and the change of function due to the change of user's behaviour is denoted by $\Delta F / \Delta x$. The model postulates that the main motive of the users is to maximise their utility function by changing their behaviour.

However, users usually have no clear quantitative description of how their actions can affect the benefit they want to maximise. In practice, it is quite likely that they will try to satisfy their motives by adapting their behaviour in a "direct search way" as described by the Hill Climbing Method in optimisation theory (Wilde and Beightler, 1967). They change their behaviour bit by bit each time they find it to be beneficial, i.e., they let $x(t + 1)= x(t) + \Delta x$ when $\Delta F / \Delta x > 0$ (see blocks 9–10).

Cognition and decision in behavioural adaptation

Once a change in the environment has been perceived, and its effects understood, the drivers have the option of deciding to change their behaviour. For behaviour adaptation towards risk change to occur, we must assume that road users' attainment of their goal requires an objective risk be taken, and the users must also perceive some level of change of subjective risk (Underwood *et al.* 1993). Block 5 and block 6 in Figure 2.2 address these issues. If the users have perceived some amount of risk change (i.e. DRs ≠ 0 in block 5), it is quite likely that they will try to take advantage of it. Could a change in the perceived risk lead to an increase in the utility function? If a potential change of the function is predicted with regard to the change of subjective risk, a further step will be taken to explore possible changes of their behaviour. Besides risk change, any change in users' goal function (denoted as ΔF) can lead to adaptation itself. We represent this factor as an independent variable input to the decision process.

For behavioural adaptation to occur, we must assume that road users have the ability to change their behaviour in response to the system changes they have perceived. Without such ability or if it has been limited by enforced traffic laws, users will have little chance to change their behaviour even when that is desired. For example, Wilson and Anderson (1980) found that drivers were able to perceive a difference between cross-ply tyres and radial-ply tyres, but did not change their behaviour very much in response to the perceived differences. Perhaps the limitations in their driving skills prevent them from taking full advantage of the change: adaptation can be initiated by introduction of more opportunities for behavioural changes but the behaviour itself must have scope for change.

These factors are represented in blocks 7 and 8. In block 8 we have used Δx to denote the set of behaviours available for adaptation. Drivers can only adapt those behaviours which are in the set, and the size of the set is determined by road-vehicle-user system. The changes in the system may lead to change of size of set, and eventually cause adaptation to occur or be stopped. A good example is the introduction of speed limits which restricts users' behaviour by legal force. This mechanism is represented by a link between block 1 and block 8.

Feedback loops in behavioural adaptation

If behavioural adaptation is to occur there is feedback to road users from the effects of their behaviour upon their immediate environment. Feedback refers to knowledge and information received from the system and which results from changes in road users' behaviour. In this sense feedback is a major component of the model and will allow drivers to modify their behaviour towards either greater or reduced safety.

The Jiang *et al.* model summarised in Figure 2.2 proposes two feedback loops. The first feedback loop contains blocks 7, 9 and 10. To allow behavioural adaptation to develop, the users need to answer the following question again and again: can the next adaptation give a good return (see block 10)? Only a positive answer to this question can convince them that their next change is rational and worth attempting. Furthermore we must assume that road users have the chance to go on to change their behaviour in response to the feedback they received (see block 7). In certain cases, drivers may be unable to take full advantage of perceived improvements in road design or vehicle performance because of limitations in other components in the system. The development of adaptation can also be limited by enforced traffic laws which control driving behaviour.

When both of the questions in blocks 10 and 7 receive a positive answer, and only at this point, the adaptation process can be repeated again and again, and the behavioural change be developed as a function of time as described by block 9. It must be recognised that the processes involved in behavioural adaptation may operate over both time and space to a large extent.

The second feedback loop in the model starts from block 1 and ends at block 1. It contains all of the numbered blocks except blocks 1 and 8. This loop describes the feedback process in which changes of the transport system lead to behavioural adaptations, and the adaptations in their turn eventually result in positive or unanticipated negative effects on transport system safety. As inputs into the loop, perception parameter a, attitude parameter b, and change of users' utility function ΔF, are regarded as independent variables which control the direction and amount of the desired adaptation. It shows that drivers adapt actions for different reasons, and on the basis of environmental demands and psychological processes.

The process of adaptation may be stopped in different stages from $t = 1$ to $t = T$ as indicated in Figure 2.2. Any negative judgement made in blocks 5, 6, 7, or 10 can lead to the conclusion that the best outcome (denoted as F^* in block 11) has been achieved in relation to the users' goal and no more adaptation is needed or available. The total change of behaviour at this stage is the sum of all changes over the period from t=1 to t=T, which is denoted by the relationship

$$\Delta x^* = \sum_{t=1}^{T} \Delta x(t)$$

which may not resemble the intended behavioural change $\Delta x'$ to be produced by initial system change. The resulting behaviour at this stage is denoted as $x(T) = x(1) + \Delta x^*$. Over a long period of time, a new behavioural habit and possibly also a new goal function will be established based on $x(T)$ and F^*, although the driver may not aware of this process (Underwood, 1982). This new habit may be a product of changed intentions or of implicit learning. The formation of the habit is represented in Figure 2.2 (block 12).

Adaptations developed over time will eventually result in certain types of effects on transport system safety. This feedback effect is indicated by a parameter of f initially proposed by Evans (1985). The feedback parameter f characterises the degree to which the behavioural adaptation influences the outcome of safety measures. The relationship between expected and actual safety benefits from changes to the system can be represented by the following formalization:

$$\Delta Ro = (1-f)\, \Delta Re$$

where ΔRo is actual change of objective risk, and ΔRe is expected change of risk.

Summary of the adaptation model

The Jiang *et al.* model of behavioural adaptation was developed to account for variations in behaviour as changes are made to the road environment with the intention of increasing safety on the roads. It can also be regarded as a model of change following the introduction of an RTI system which requires a change in user behaviour. The model suggests that:

- The behaviour of drivers cannot be considered to remain constant.

- Drivers adapt their behaviour to changes in their surroundings, following a tendency to satisfy their goals.

- The adaptation mechanisms of drivers can operate at different levels and develop through feedback processes as a function of time.

- Drivers do not necessarily have a conscious evaluation of their risk of accident involvement in their behavioural adaptation, more likely, they depend on a feeling of risk change.

- Drivers do not necessarily have a quantitative and consistent representation of their goal or benefit—awareness of their improvement is enough for directing the adaptation.

- Following changes in the traffic system, drivers may or may not detect associated changes in objective risk.

- Drivers can be discouraged or prevented from behavioural adaptation.

- Drivers can only partially compensate for changes in objective risk.

- The behaviour prediction or explanation provided by the model applies at the individual level.

The model acknowledges that behavioural adaptation is a pervasive phenomenon in traffic systems. In some cases it may even generate effects opposite to those intended. The model describes the interactive and feedback nature of the adaptation process as represented by the closed-loop flow chart. It explains the process and indicates where and when behaviour adaptation is likely to take place and its potential effects on safety measures. The model should not be seen simply as a method of assessing the effects of behavioural adaptation. In the long term, positive and sustained benefits can be expected if a better understanding of the mechanism of behavioural adaptation leads to a better formulation and implementation of safety measures. Only by taking account of the changes in driving behaviour that accompany specific engineering changes will we be able to achieve our aim of developing RTI systems which maximise the safety of the road transport system.

SUMMARY

The short-term solution to present road traffic problems is to utilise existing road networks more efficiently. One way to achieve this is the implementation of information technology in road transport management: RTI. Over time RTI systems will supersede, though not necessarily make redundant, traditional forms of information gathering and dissemination on road networks.

The degree to which an Integrated Road Transport Environment (IRTE) can be achieved, depends on factors such as the usability characteristics of the system used; the proportion of drivers equipped with the system in question; and the extent to which drivers act upon the advice given by the system. We will only achieve our aim of developing a more efficient and safer road transport system if RTI products not only serve their primary function of communicating information on a wide scale, but also are designed around the performance characteristics of the prospective users. The most effective systems will be those designed around the needs and abilities of the drivers who will operate them. We need to take account of the human factors associated with the effective operation of developing information systems and those associated with behavioural adaptation to change.

An RTI application area is fulfilled by an RTI system. This indicates that the application specifies to *what* use RTI can be put in a particular locality, and the system specifies *how* this can be achieved. The product is the physical deliverable which implements the RTI system. The potential range of RTI applications and systems is large. Thus in the diversity of possible options, a methodological decision process should be imposed in order to choose the optimal solutions for a particular locality. Such a method, introduced in Section 2, is the assessment procedure.

Section 2

THE ASSESSMENT PROCEDURE

3 OVERVIEW OF THE ASSESSMENT PROCEDURE

We start with an overview of a pragmatic approach to the evaluation of RTI applications in a particular context. Figure 3.1 illustrates the assessment approach, which combines top-down analysis of RTI applications with bottom-up synthesis of RTI technologies. Assessment can be seen as the process of linking theory, which in the present case is a physical model, with its realisation as a complete RTI system. Assessment is necessarily iterative and cyclical, being performed continuously and depending on both the state of RTI system development and of analysis of known RTI applications. This text does not emphasise the mechanics of modifying and redesigning systems, and therefore the procedure appears more linear than it is in reality. It will make use of laboratory tests, prototype tests and field tests as well as expert opinion and field surveys in assessing the RTI systems that are most

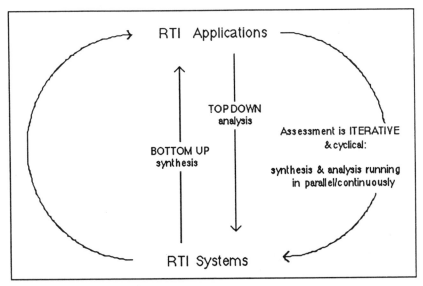

Figure 3.1 General assessment approach

appropriate in a particular locality. The assessment procedure is based upon
that proposed by the DACAR project, and is elaborated in Altendorf *et al.*
(1990, 1991).

The procedure for assessment can be broken down into three sub component loops (Figure 3.2). Each loop ends with a command to list the systems
which meet the requirements specified at the decision point feeding into it.

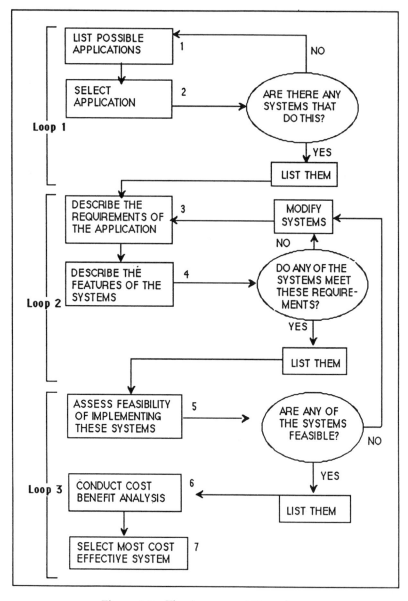

Figure 3.2   The Assessment Procedure

The structural flow diagram will act as the reference point throughout Section 2. It depicts the key stages using blocks 1–7, and key decision points using oval shapes.

The assessment procedure consists of three main loops, each loop ending with a request to list RTI systems. The procedure starts with the choice of application. This is accomplished by first listing all possible applications (Block 1 in Figure 3.2), then selecting the most appropriate for the particular locality in which it will be implemented (Block 2). The process of matching systems with applications begins with establishing the abstract relationship between systems and applications, illustrated by the physical model (see Figure 3.7). The first decision point is then tackled: "Are there any systems that fulfil this application?". If there are, these should be listed.

Block 3 marks the beginning of the second loop. The requirements for RTI applications and systems are described in two stages: stage 1 matches systems with the physical model of the application, then stage 2 matches the system with the technical requirements. Once these requirements have been described, the second decision point is reached: "Do any of the systems meet these requirements?" If the answer is "yes" they should be listed. The next phase of the selection process assesses the feasibility of the most appropriate RTI systems from this list (Block 5). The final stage is to conduct a cost benefit analysis (Block 6) of feasible systems.

Assessment and evaluative procedures are generically iterative. Despite this, we have assumed that the RTI assessment procedure begins with a top-down analysis; selecting the appropriate application(s) for a given locality. Although this starting point will be the one most commonly experienced, the procedure can be altered with moderate ease to accommodate an alternative starting point.

CHOOSING AN APPLICATION

Table 3.1 contains an extensive breakdown of RTI applications. Several other ways of listing RTI applications exist, as well as other ways of grouping applications. However, for the assessment procedure we will work with the more manageable classification of seven groups proposed within the DRIVE framework. These groups were reviewed in Chapter 1.

- Demand management;

- Traffic and travel information;

- Integrated urban traffic management;

- Integrated inter-urban traffic management;

- Driver assistance and co-operative driving;

Table 3.1   A classification of applications

Function group	Functional area	Applications
Driving support systems	Cooperative driving systems	1.1 Intelligent manoeuvring systems 1.2 medium-range preinformation systems 1.3 data collection for traffic management 1.4 intersection control systems 1.5 automatically guided systems
Traffic control systems	Route guidance systems	2.1 actualisation of autonomous route guidance systems 2.2 dynamic route guidance, traffic depentent, locally 2.2 dynamic route guidance, traffic depentent,regionally 2.4 long-distance route guidance during holidays
	Parking info systems	3.1 operation of parking gates and garage doors 3.2 parking management, park and ride systems 3.3 booking of parking spaces for passenger cars 3.4 booking of parking spaces for trucks (transit traffic) 3.5 booking of reserved space for just in time delivery
	Driver info systems	4.1 local DIS on road works, detours etc. 4.2 regional DIS on road works, detours etc. 4.3 local non-traffic-oriented DIS 4.4 regional non-traffic-oriented DIS
	Warning system	5.1 transmission of warning signals into the car 5.2 warning of accidents and incidents 5.3 emergency call systems
	Traffic mgt info system	6.1 optimisation of traffic lights at single intersections 6.2 local coordination of traffic lights 6.3 traffic management in cities 6.4 regional traffic management
		7 tunnel management systems
Special services	Commercial & public fleet mgt	8.1 transmission of data from and into vehicles 8.2 public transport guidance systems 8.3 regional fleet management (taxis, police) 8.4 long-distance fleet mgt (truck fleets, just in time del.)
	Automatic debiting systems	9.1 automatic fee collection systems on freeways 9.2 automatic fee collection systems of parking lots 9.3 roadpricing 9.4 vehicle taxation on the territorial principle

Note: This grouping of functions should not imply a "tree" structure. A degree of overlap is present, where an application may belong to more than one function.

• Freight and fleet management;

• Public transport management.

When selecting the most appropriate RTI application, the extent to which the application meets the agreed objectives must be considered, as well as matching the description of the application with the needs of the locality.

An influential technical report by von Tomkewitsch and Kossack (1989) advocates that applications be classified according to their range of action and the required reaction time (see also Figure 3.12). Thus *four* system groups help to refine the application selection:

Group A   Applications with an *isolated* range of action   (< 100 m)

Group B   Applications with a *local* range of action   (< 5 km)

Group C   Applications with a *regional* range of action   (< 25 km)

Group D   Applications with a *transregional* range of action   (> 25 km)

For example, automatic toll debiting would generally be classed as an application with an isolated range of action. Parking management, depending on contextual factors, could be classed as having a local or a regional action range. Traffic management/route guidance in an urban environment would usually have a regional range of action, and fleet management would generally be classed under group D, supra or transregional. There are numerous applications which could be seen to have an extensive range of action, and thus could not be classified under one or more group headings.

From the initial list of applications, then, the choice will be limited by the desired range of action. If more than one application can fulfil the initial selection criteria of meeting traffic goals across an appropriate spatial level, then additional criteria will need to be applied. For example, each application can now be rated on the basis of its contribution toward traffic safety, the estimated effectiveness and the chances of realisation (von Tomkewitsch and Kossack, 1989).

In general, the reasons for introducing RTI are to improve traffic safety, to reduce pollution, to improve traffic efficiency, but traveller comfort is often of importance. Evaluating a particular application should take each of these factors into account. They should be differentially weighted since, for example, traffic safety in general has a greater priority than driver comfort. The following weightings have now become generally accepted:

Traffic safety   3

Traffic efficiency   2

Environmental protection 2

Driver comfort   1

Each possible application should be given a graded score, generally between 1 and 5, for how well it achieves each of these objectives. To calculate a contribution rating for each application, multiply those scores by the weightings given above before summing them.

The effectiveness of each application is multiplied by the contribution scores:

If the number of affected persons is	*Or* the expected benefit is	*Then* the traffic policy priority is	multiply by:
low	low	low	W = 1
medium	medium	medium	W = 2
high	high	high	W = 3

Irrespective of the contribution and effectiveness of an application, the probability of realisation must be taken into consideration. The chances of realisation can be graded from 0 to 3 (unrealisable to highly probable). Further discussion of this issue will take place in Chapter 4. Ideally, the scores given to each application on each factor will be based upon existing research. When information on the effectiveness of the application is unavailable, the assessment will need to be based upon individual, expert opinion. These weightings are largely subjective and may well vary with, for example, political climate. For example, with increasing public outcry, environmental protection (EP) will become increasingly significant and will be awarded a higher weighting factor. European integration will also influence the way in which individual member states prioritise RTI applications.

In order to select the most appropriate application for a given locality, taking into account local traffic needs, for example, applications should be assessed according to predefined criteria. Each application can then be subjected to an evaluation according to the decision analysis process, proposed by von Tomkewitsch and Kossack (1989). A hazard report application would be classified as a warning transmission which could be frequently applied only at one specific position (action range Group A) or region (action range Group B). Such an application would be given a high score on the traffic safety weighting. A detailed evaluation of a possible congested town scenario is shown in Table 3.2. The suggestions and calculations are far from definitive, and the numbers will change for different types of cities and regions. A solution that is optimal for one region may be highly undesirable for another. The evaluation scheme has to be operated with great care and to whom the selection applies should be taken into account.

Table 3.2   Applications for a congested town scenario

Some suggested applications to resolve town congestion	TS	EP	TE	DC	W	R
	3	2	2	1		
Local range traffic management						
External route guidance	1	2	2	3	14	2
Autonomous routing	1	1	1	3	10	2
Parking management and park and ride information		3	3	3	15	3
Tourist information						
Isolated range/static information (e.g. hotels)		1	2	3	9	2
Local range/dynamic information (e.g. free hotel rooms)		1	2	3	9	3

TS = traffic safety   W = effectiveness
EP = environment protection   R = probability of realisation
TE = traffic efficiency
DC = driver comfort

MATCHING SYSTEMS WITH APPLICATIONS

If the procedure were to be followed, we would now be at the stage in the assessment procedure where a decision about application selection will have been made. Here we outline a decision process for identifying those systems that will, from both a theoretical and a technical point of view, fulfil this application (Blocks 3 and 4 and accompanying Decision Point in the assessment procedure). Figure 3.3 illustrates the relationship between application, system, and product.

The technical requirements of the application "motorway toll automatic debiting system" (block 3), will be discussed in Chapter 4.

Once a physical model for the application has been developed, it is possible to think about how existing systems and technologies might be used to implement it. Here we provide some suggestions that will provide a guide to the variety of systems/products on the market or under development, and which are available to fulfil specific applications. It must be emphasised that this list is far from exhaustive, rather it offers a selection of the total range and, because RTI is a new and expanding field, these products will also date very quickly. Current information can be obtained from the annual reports produced by manufacturers or distributors, or from research organisations such as universities and specialised consultants.

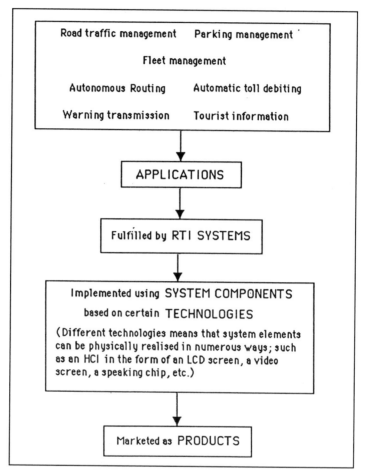

Figure 3.3   The relationship between system and application

Existing systems

We can now illustrate a range of RTI systems, applications and products on the market or under development, using the DRIVE-defined applications areas. These are summarised in Table 3.3.

To further illustrate approaches that have been used, we can consider a sample of existing products mentioned in Table 3.3:

- CARIN/SOCRATES
- PRODAT
- MCSS
- Autoguide.

Table 3.3   A list of existing RTI systems, applications and products

RTI system	RTI product	System components	RTI applications	Manu-facturer
RDS-TMC Radio systems GSM Radio systems			C:Warning systems, A: driver information A: driver information F: commercial & public transport management	
Radio systems, microwave & infra-red			A: Route guidance (IRG)	
On-board navigation system	CARIN (uses a location display)	*Sensors* (for speed & direction *information processing* (CD ROM & small computer), *HCI* (digital map map, keyboard & speech synthesiser)	B: Route Guidance, driver information	Philips
On-board navigation system	SOCRATES (not yet a commercial system)	CARIN + GSM	A / B: Dynamic Route Guidance	Philips (one of the consortium)
	MCSS	*Sensors* (traditional inductive loop), *information processing* (central computer, local processors and terminals)	A: Road traffic control	Dutch RWS
	Prodat	*Sensors* (satellite receiver), *Information processing*(RAM storage,small computer) *HCI* (printer, keyboard) *Information storage & processing* (infra-red beacons, local traffic control centre)	A: Communication between vehicles & parts of the infrastructure	Siemens-Plessey

Table 3.3 *(continued)*

RTI system	RTI product	System components	RTI applications	Manu-facturer
Externally linked route guidance system	Autoguide (not a commercial system)		A: Route guidance in urban environments	Siemens Plessey/ GEC
Automatic vehicle ID (AVI)	ID-tag PREMID KOFRI		A: Road pricing (a tool for traffic restraint policy) D: Automatic debiting (UFFI)	AMTECH Philips Micro-design
Electronic route planning	TELETEL videotext	comprises many interactive services	B: Pretrip route planning	French
Externally linked route guidance systems	ALI-SCOUT (infra-red based)		A: Route guidance	Bosch/ Blaupunkt & Siemens
	TravelPilot		B: Autonomous route guidance	Blaupunkt
	GSM		Digital mobile telephones	SEL. Nokia etc.
	GPS		Localisation system	SEL.
Electronic route planning	ROAD-WATCH	*Information processing & storage* on a database in infrastructure, updated & provides traffic information.	B: Pretrip route planning	AA (UK)
Traffic information broadcasting		Transmits traffic information using a subsidiary FM receiver (no manual tuning required)	A: Road traffic management	FHWA (Federal Highway Administ-ration) US

Table 3.3 *(continued)*

RTI system	RTI product	System components	RTI applications	Manu- facturer
Automatic vehicle location (AVL)	GEC TRACKER system	*Sensors* (dead reckoning + updates from proximity beacons), information processing (linked to control centre).	F: Fleet Management	GEC (UK)

Key:
A   Road traffic management   E   Parking management
B   Autonomous routing   F   Fleet management
C   Warning transmission   G   Tourist information
D   Automatic toll debiting

For additional information concerning cable, microwave and infra-red systems, refer to Section 3.

The CARIN/SOCRATES System

A notable example of an onboard navigation system, the CARIN system, is not a dynamic route guidance system; it is a read-only CD-ROM data storage system which does not take into account the current road and traffic situation. To overcome this, the SOCRATES consortium has designed a system that makes use of a mobile radio to actualise the information stored on the CD-ROM (Figure 3.4). A thorough description of the SOCRATES project is given by Catling *et al.* (1991). This involves adding components to the CARIN system which enable it to receive the information on the new pan-European cellular radio system (GSM). Ott (1977) gives a fuller description of cellular radio systems. For further discussion concerning dynamic route guidance, see the glossary and also Chapter 5.

The PRODAT system

In this system the HCI is a small printer and keyboard. The driver can type messages that will, after a specific command, be transmitted to a centre such as the company's main office. The processor and the storage system can select, decode and store received messages and then code outward messages. The communication system consists of a satellite transceiver that can

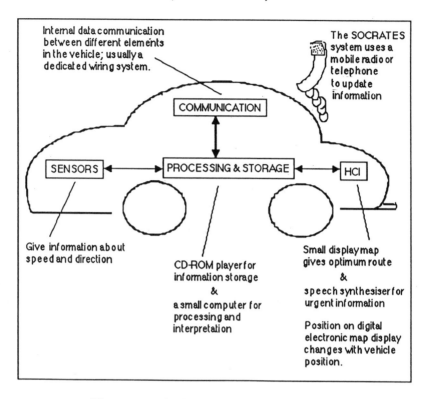

Internal data communication between different elements in the vehicle; usually a dedicated wiring system.

The SOCRATES system uses a mobile radio or telephone to update information

COMMUNICATION

SENSORS ◄───► PROCESSING & STORAGE ◄───► HCI

Give information about speed and direction

CD-ROM player for information storage & a small computer for processing and interpretation

Small display map gives optimum route & speech synthesiser for urgent information

Position on digital electronic map display changes with vehicle position.

Figure 3.4   The CARIN / SOCRATES system

send and receive messages via a satellite. Because the system does not have any sensors, the location of the vehicle needs to be typed in by the driver. The complete in-vehicle equipment costs approximately 5000 ECU. In addition there are the communication costs of 1 ECU per kilobit per second. The system is considered to be useful for European transport companies and logistic service providers, as well as for servicing public transport and police communications.

The MCSS management system (Motorway Control and Signalling System)

Commissioned by the Dutch Rijkswaterstaat (Ministry of Water, Road and Transport), the MCCS infrastructure system (Figure 3.5) is based on approximately 80 measuring points in the form of sensors that identify the number of vehicles, their speed and their type. These sensors are connected to a local intelligent roadside processor, which in turn is connected to a central computer

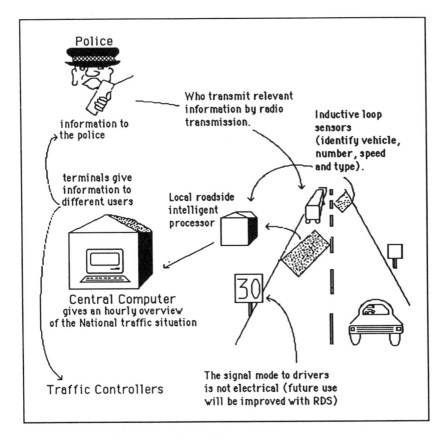

Police

Who transmit relevant
information by radio
transmission.

information to
the police

Inductive loop
sensors
(identify vehicle,
number, speed
and type).

terminals give
information to
different users

Local roadside
intelligent
processor

Central Computer
gives an hourly overview
of the National traffic situation

Traffic Controllers

The signal mode to drivers
is not electrical (future use
will be improved with RDS)

Figure 3.5 The MCSS management system

that is able to give an overview of the national traffic situation on an
hourly basis. This information is also given to the police, who are then able
to broadcast congestion information. The monitoring network does not communicate with the drivers, it is an information collecting agent only. Driver
communication and traffic control are implemented through automated
roadside signs. A new version of the monitoring network is being developed which will improve the system's performance, by connecting it to
other types of systems. In particular, use will be made of the traffic message
channel on the Radio Data System (RDS-TMC), so that the information that
is gathered will be transmitted directly as digitally encoded messages on
RDS. This makes it possible to inform drivers in such a way that they can
take into account the actual traffic and road situation. The price of the complete new network will be approximately 20 million ECU. For elaboration of
RDS-TMC, see Chapter 6.

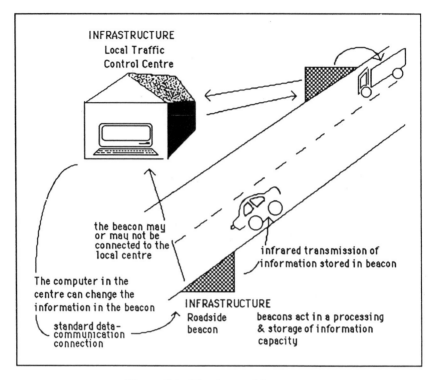

Figure 3.6   The Autoguide system

The Autoguide route guidance system

Catling (1987) and Jeffrey *et al.* (1987) have described this system in detail. Originally there were no sensors in the Autoguide infrastructure (Figure 3.6). Autoguide deals only with the communication component of the application. In this case, the infrastructure consists of beacons that may (or may not) be connected to a local traffic control centre. The communication system consists of an infra-red beacon that transmits stored information on routes taken by individual drivers in a certain part of a town. The information contained in this beacon may be changed by connecting the beacon with a standard data communication connection (e.g. telephone lines with modems) to a computer in a traffic control centre. For further information concerning the infra-red system, see Section 3.

SYSTEMS THAT CLAIM TO FULFIL THE APPLICATION NEED

In this chapter we have discussed the relationship between systems and applications. We now arrive at the first decision point in the assessment procedure (Figure 3.2, Loop 1): "Are there any systems that claim to support the application?"

Identifying appropriate systems requires a careful description of the intended application. So far we have identified the general function that the RTI system will fulfil, and the system's possible range of action. It is now necessary to define it in considerably more detail. This needs to be done in two stages. Stage one entails matching systems with the physical model of the application. Stage two then matches systems with the technical requirements of the application.

Matching the system with the physical model of the application

The first stage requires the specification (in a non-technical way), of the system components that will be necessary to fulfil the application. This needs careful consideration of the nature of the locality in which the system will operate.

Assume, for example, that the chosen application is parking management. Parking management needs vary considerably from town to town. To produce a full description of the system that will fulfil this application it will be necessary to decide how often the information supplied to drivers will need updating; whether information on-street parking (as well as car park) spaces need providing; how far from the town or city centre is the range necessary for drivers to receive this information; and a whole range of similar details. On the basis of these it should be possible to decide on the system components, and the direction of information flow between these components, that will be required to fulfil the application. This non-technical description of the system we will call the *physical model*.

Figure 3.7 suggests possible physical models for systems fulfilling RTI applications in the seven groups that we have already identified. These are, of course, only possible solutions; we have already stated that any one application can be fulfilled in a number of different ways. In the model we have used *active, passive* and *responsive* to mean the following:

- *Active:* Gathers information in a continuous manner from the control system; advertises itself as a sensor and invites a response.

- *Passive:* Gathers information in a non-continuous manner. There is access only to a non-selective quality of information.

- *Responsive:* Responds to information gathered.

The decision procedure in Figure 3.8 suggests a mechanism for identifying suitable system elements for the applications that are being considered. It is essentially very straightforward, and starts with the specification of a

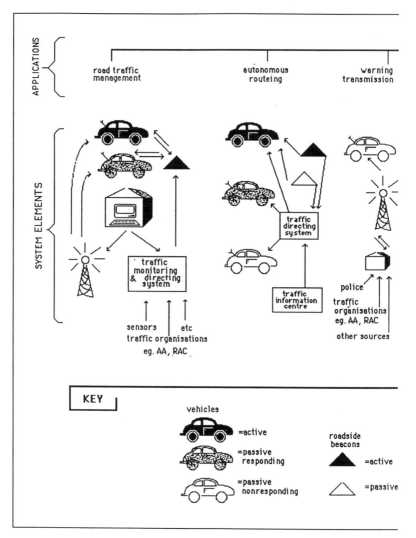

Figure 3.7   The physical model

physical model for the application. For instance, suppose the required application falls within the road traffic management category: the system elements required would include the following:

- Active vehicle
- Passive vehicle
- Active beacon
- Traffic control centre

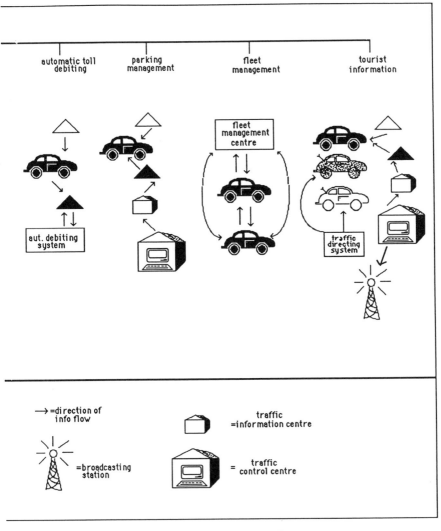

automatic toll
debiting

parking
management

fleet
management

tourist
information

fleet
management
centre

aut. debiting
system

traffic
directing
system

→ =direction of
info flow

=broadcasting
station

traffic
=information centre

= traffic
control centre

Figure 3.7 (*continued*)

- Broadcasting station
- Traffic monitoring and directing system.

Matching the system with the technical requirements of the application: the OSI architecture

The seven-layer ISO-OSI (International Standards Organisation–Open Systems Interconnection) Reference Model is a theoretical model originally

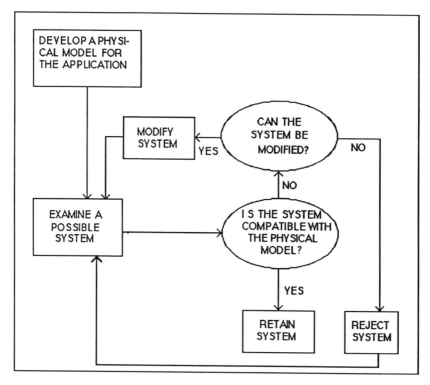

Figure 3.8   Matching systems with a physical model

designed to standardise the procedures and communication process for interconnecting autonomous computers grouped in a network. It can be used to illustrate the technical levels, and thus requirements, of RTI applications because:

(1) The OSI model allows analysis of different types of communication systems (from simple broadcasting; point-to-point; selective networks; to co-operation between mobile changing partners).

   The following description assumes that even the most sophisticated type of communication between mobile changing partners, necessitating the use of a dynamic communication network, can be described. In reality, the context will probably entail a dynamic network formed by several vehicles, roadside infrastructures, central computers etc., which co-operate actively to exchange relevant information. Simpler systems will also be accounted for.

(2) The OSI model is systematic in describing all aspects of the necessary RTI infrastructures.

(3) The OSI model provides the possibility of discussing technical transmission issues as well as those issues, related to contents of the

transmission, in an orderly manner. As it is supported by a set of international standards, this provides possibilities for relatively fast system integration.

General features of the ISO-OSI model include:

- A description of a seven-layered set of interfaces, necessary to build a dynamic communication network.
- Each layer provides services to the layer above and receives services from the layer below.
- The layered set of interfaces means the possibility exists to design and modify each layer independently, provided the interfaces remain unchanged.
- The *n*th layer of one machine can converse with the corresponding *n*th layer of another using rules called protocols.

The seven layers can be categorised as being either *application-oriented* or *transport-oriented* layers, and are represented together in Figure 3.9. The relationships of the layers can be summarised as follows:

Layer 1 In the Physical Layer of the communication link, raw data (bits) are transferred between two neighbouring machines. It is at this level that physical communication occurs between machines.

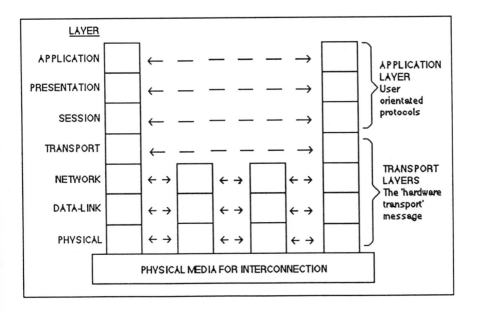

Figure 3.9   The OSI architecture

Layer 2 The Data Link Layer supplements the raw data transfer by performing error control, thus transmitting error-free data to the higher layers. As Layer 1 exchanges bit streams regardless of meaning or structure, Layer 2 ensures a reliable link quality by means of certain techniques.

Layer 3 The Network Layer supports multiple links on a single transmission path, (thus enabling addressing features and including the call control functions). Therefore the Network Layer provides a network connection path between a pair of transport entities.

Layer 4 The Transport Layer provides an end-to-end transport service: the main and important difference between Layer 3 and Layer 4 is that Layer 4 has a more general addressing scheme and enables the information exchange among a number of different networks. The sender and receiver are unaware of the specific routes taken by the data.

Layer 5 The Session Layer enables a logical connection which is continuous and appropriate, by defining the structure of the dialogue (usually called session) between two machines.

Layer 6 The Presentation Layer deals with the transformation of the received data according to user-format, also providing data code conversion when necessary. Layers 6 and 7 determine what is displayed, and how it is displayed.

Layer 7 The Application Layer entails the response of individual users (or an application), who determine the action to be taken upon receipt of each allowed message.

The ISO-OSI reference model distinguishes between the Transport ("hardware transport" layers 1–4) and application ("user oriented" layers 5–7) layers. It is therefore possible to restrict further analysis to the lowest layers which are affected by the transmission medium, and consequently by transmission techniques. For the higher layers the discussion should focus upon individual applications. In the following we assume that a certain application is chosen and that a remaining task is to further define the lower layers. If the ensuing description used anything more complex than a single network then we also have to take into account the overall structure of the IRTE. The emphasis here will be on mobile communications, which are of central importance for the implementation of RTI.

After noting the components of a generic communication link, we will present the technical requirements for OSI levels one to three. For the purpose of technical requirements in RTI, we will assume that level one of the OSI model corresponds to the message level, level two to the transmission level, and level three to the network level. For further description of the ISO-OSI model as applied to RTI, see, for example, Wall and Williams (1991) and Bueno and Ongaro (1991).

COMPONENTS OF A COMMUNICATION LINK

The main functions and subsystems included in a transmission system or communication link are summarised in Figure 3.10. This link could be

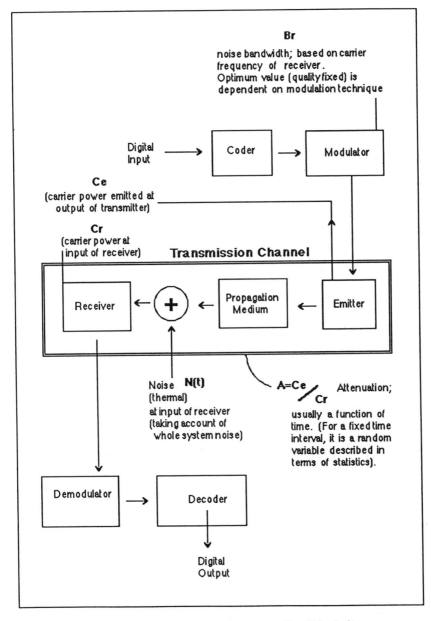

Figure 3.10 A communication link: a generalised block diagram

between any two elements in an RTI system, for example, between a roadside camera and a local information centre, or between a vehicle and a roadside beacon. The notation given in the figure is also used in the sections that follow. In order to make an analysis according to Figure 3.10, note that the purpose of the data transmission system is the transmission at a given quality of the bit stream I, with a particular structure, where the information bit rate associated with input I, is denoted B_{ri}.

Suitable coding can be used in order to achieve both larger robustness to transmission impairments and good spectrum efficiency. In order to transmit the information through the link, a suitable modulation technique has to be used, which generally exploits a carrier frequency f, suitably allocated in one of the available frequency bands in the electromagnetic spectrum.

There is potential either for information to be lost or for noise to be introduced at each stage in the communication link. This introduces the possibility of communication errors (i.e. the information that is received may be either incomplete or different from the information that was transmitted). One measure of the extent to which a communication link introduces errors is the "bit error probability", which is defined as the probability that, at any given instant, the information leaving the demodulator will be different from the information that entered the coder (often denoted as P_e). The performance of the communication link is thus a function of:

- how it codes information;

- the modulation and demodulation methods;

- the nature of the propagation medium.

For a particular communication link to be successful, it must transmit information with a certain minimum accuracy. This will be dependent upon its function within the context of a whole RTI system, and may be expressed in terms of a minimum bit probability error.

Transmission system components include:

- *Coder:* which codes the information that is to be transmitted (e.g. a roadside camera might translate what it sees into binary code).

- *Modulator:* turns the coded information into a signal that can be transmitted (e.g. from binary code to a radio frequency wave).

- *Transmitter:* transmits the signal (e.g., an antenna, an infra-red diode, or a microstrip).

- *Propagation medium:* is the medium through which the signal will travel (e.g. in the case of radio waves, the medium is air, and in the case of light it is optical fibre).

- *Receiver:* receives the signal.
- *Demodulator:* turns the signal (such as radio frequency) that was transmitted, back into code that can be read by the decoder.
- *Decoder:* turns the coded information into a form that can be more readily used (e.g. from binary code to a visual display at the local information centre).

We will 3now discuss the main performance requirements at the message, transmission and network levels.

Requirements at the message level

The message structure will be presented first, followed by an outline of the parameters. Information is represented in the physical layer by a digital message. The message consists of a finite number of information bits, n. There are also additional bits in the data packet, such as start and stop bits. The length of the packet, t, is determined by the application. As an example, in a roadside-to-vehicle transmission, the packet length is chosen according to the maximum vehicle speed and the beam width of the roadside beacon. The value of T represents the repetition time for information which does not vary (transmission can employ a random multiple access protocol in a selective communication network, when packet re-transmission is not periodic).

Minimum information bit rate

This is a measure of the mere information amount, which is evaluated with consideration to the end-user throughput, irrespective of retransmissions etc. occurring. The information bit rate B_{ri}, can be expressed as information amount over a fixed data packet length: $B_{ri} = n/t$.

Maximum packet repetition time

This information is not necessary in all applications. When the information to be transmitted does not change with time (e.g. automatic debiting), a repetition of the same message is not strictly required, but can be justified for security or redundancy purposes. In most RTI applications a refreshment is requested due to the variation with time of the information contents. Therefore, on such occasions, this parameter would be valued according to the rapidity of the related variations.

Repetition of information is a useful way of reducing errors (Figure 3.11), and the information content plus the redundancy introduced through repetition increases the quality and reliability of transmission $= B_r$ (transmission bit-rate). This increased transmission quantity is represented by

$$B_r > B_{ri}.$$

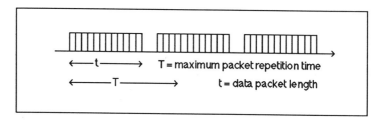

Figure 3.11   Repetitive packets

RTI applications are in general characterised by "burst-mode transmission". However, in some particular cases "continuous flow" data transmission can be required. This can be considered as a particular case of the previous one, by putting the repetition time equal to the packet length, t.

Requirements at the transmission level

At this level we distinguish, as the main performance parameters, the minimum range of transmission, the maximum vehicle speed, the maximum bit error probability, and outage.

Minimum range of transmission link

This parameter usually depends on some architectural aspect and is not unequivocally determined by the application. Some RTI applications require short-range links, such as the application based on vehicle beacon communication. In other circumstances, "link range" is not an appropriate parameter. Achieving a particular application could entail exploiting systems already in existence or modifying/redesigning systems. For example, when general traffic information is extended to regional areas, the existing infrastructure (such as RDS or GSM) could be exploited to transmit this information. In such a case, the minimum range of transmission link is no longer a sensible parameter.

Applications can be categorised into four classes depending on their range of action, as illustrated in Figure 3.12. Applications utilising an isolated range of action require minimal system elements such as the vehicle and roadside beacon. The local action range may also need a local computer and broadcasting station. The regional range category could require these elements plus a traffic management /control centre. In addition to all the system elements above, those applications which extend beyond a region would include use of a mainframe computer in central control. This last category of action range could well entail a situation where minimum range of transmission link is no longer a constricting parameter. For further information concerning link range, see von Tomkewitsch and Kossack (1989), and specifically Figures 1 and 2.

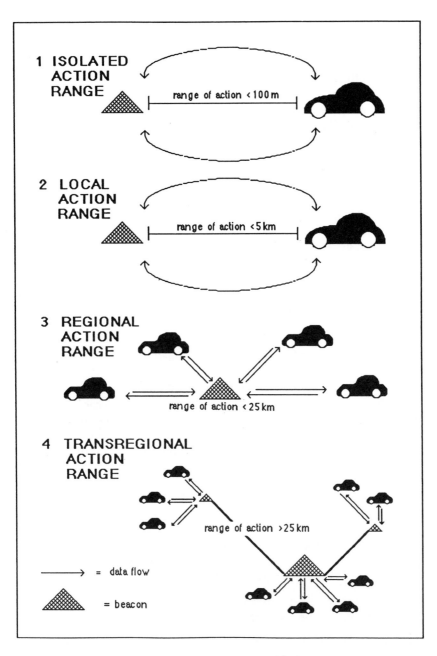

Figure 3.12   RTI applications and link range

Maximum vehicle speed

This constraint of information flow affects both the propagation mechanisms and the maximum time interval.

(1) *Propagation mechanisms.* The multipath signals of a typical mobile channel are affected by the relative speed of the two transmitter sites. The characteristics of the signals change as the speed varies. Such channel variations must be compensated using algorithms, to ensure a good quality link. The algorithms entail the adaptive equalisation necessary for this link, and are related to the transmitted bit rate and the maximum vehicle speed.

(2) *Maximum time interval.* The time interval when the receiver antenna is properly illuminated by the transmitting one is inversely proportional to the maximum relative vehicle speed between communicating terminals. This is a more intuitive aspect, which basically means that as the vehicle goes faster, there is less opportunity for transmission. For example, in the vehicle-to-beacon communication, the maximum range is fixed. Therefore, the total information exchange decreases as the vehicle speed increases. In order to preserve total information amount, the bit rate must also increase.

Maximum bit error probability

This is the *a priori* measurement of the greatest likelihood of error occurring during the bit regeneration process. The bit probability error in a particular communication link shows how the signal is affected in a data transmission system, by noise or fading. Such a probability depends upon a number of characteristic parameters, such as the signal-to-noise ratio at the receiver. Bit error probability can be calculated as follows:

(1) C_e = carrier emitted power at the output of the transmitter.

(2) C_r = carrier received power at the input of the receiver.

(3) $A = C_e/C_r$ = attenuation due to the propagation medium. The attenuation is generally a function of time and, for a fixed time interval, is a random variable to be suitably statistically described.

(4) $N(t)$ = equivalent additive thermal noise at the input of the receiver, taking into account the noise of the whole system.

(5) B_r = receiver noise band width (centred around the carrier frequency) of the receiver. Its optimum value, for fixed quality, depends on the modulation technique used.

(6) Signal-to-noise ratio $SNR = C_r/N_s$ where N_s is the equivalent noise power at the input of the receiver, based on the bandwidth B.

(7) Intrinsic signal-to-noise-ratio $W = C_r/N_b$ where N_b is the noise power at the input of the receiver, evaluated over a conventional bandwidth, numerically equal to B_r.

(8) Bit error probability P_e = probability that the regenerator included in the demodulator makes an error in a decision instant.

Once all this is defined, the generic behaviour of a digital transmission system can be described by the following function:

$$P_e : f(W)$$

This corresponds to a particular choice of the modulation and coding methods. If a quality performance level given by a maximum value for bit error probability is defined as P_{eo}, a minimum value for intrinsic signal-to-noise ratio is required in order to assure the desired quality. Then, if the power spectral density of the noise N_o, is known, a suitable value of the transmitted power can be estimated as a function of P_{eo}, if the attenuation of the propagation medium is also known .

Finally, it should be underlined that the consideration of the different parameters involved in the link budget (signal processing, modulation complexity, emitted power, antennae, noise figure of the receiver) also lead to some constraints for the implementation of the physical devices realising the various subsystems.

The probability can be reduced by means of countermeasures such as coding techniques, interleaving, diversity schemes and equalisation. It should be noted, however, that all these improvements result in an increase in system complexity.

Outage

Outage can be defined as the probability that the bit error rate is greater than a fixed value. In the design of a digital, wireless transmission system, particular attention has to be paid to the link behaviour. This depends on the distance between the transmitter and the receiver, and generally behaves in a random way due to various causes (such as scattering, atmosphere absorption, diffraction due to obstacles). Thus "outage" estimates the random nature of the propagation medium: "system availability", defined as the probability P_{out} that the bit-error rate P_e, is greater than a fixed value P_{eo}:

$$P_{out} = \text{Prob} (P_e > P_{eo})$$

In terms of technical requirements for applications, these are related to the safety margins determined by a specific application, then "weighted" according to generally accepted priorities. This weighting can affect the outage. For instance, according to the DRIVE priorities, driving comfort is less important than traffic safety, making the requirement on outage more stringent.

Requirements at the network level

Maximum number .of users in the network

Depending on the particular application, either the number of users is imposed as an unmodifiable input, or an upper limit can be stated. For instance, in a co-operative driving system, the number of vehicles involved in the communication process is not negotiable. Therefore, although system designers can use different network topologies and resources, they cannot alter the number of vehicles that have access to that communication network. For other applications, however (such as parking reservation systems or automatic debiting), designers can decide upon the upper limit for the number of simultaneous users.

Maximum end-to-end packet delay

This is a measure of the efficiency of the multiple access scheme adopted. If the technique selected implies some possible collisions among transmitted packets (requiring retransmission of the same packet), this results in a delay in the delivery of the information content. When the maximum delay value is exceeded, the information packet is lost. The corresponding performance degradation can be controlled according to the priorities mentioned above. There is flexibility as to the importance of this requirement, which again depends upon the particular application in question.

If the number of users simultaneously served in a given area is M, and the number of communication channels within the frequency band available for a given application is N, then the ratio M/N can be a useful parameter to characterise the spectrum efficiency from the point of view of channel management. This provides a suitable parameter to summarise the efficiency of the multiple access strategy chosen, and must in general be evaluated by taking into account not only the possible collisions among different packets, but also the probability of making an error in the packet regeneration process. Several transmission impairments, such as linear and non-linear distortion (i.e. fading, interference and noise), are general features which must be considered in the evaluation of transmission quality.

A worked example of the technical requirements for the application of a motorway toll automatic debiting system will be given in Chapter 4.

SYSTEM CHARACTERISTICS

We can now summarise the different steps which have to be taken in order to understand how a data-transmission system performs.

For a particular application characterised by a given range, and given information bit range, suitable quality and availability levels have to be established. Then a particular technique can be chosen regarding the "way to implement that application".

When the set of parameters that can meet the requirements is stated, one has to estimate the subsystem implementability and accompanying cost.

For the sake of conciseness, the main parameters describing the system behaviour have been summarised in Tables 3.4 to 3.6. They summarise the main parameters concerned with the performance level required and the system characteristics.

These tables, filled out for different techniques, allow a fast comparison among different ways of implementing the data transmission system for a particular set of RTI applications.

Comparing application requirements and system performance

We now come to Decision Point 2 (Figure 3.2, Loop 2) in the *Assessment Procedure*: "Do any of the systems meet these requirements?"

Table 3.4   Types of communication

(1) Vehicle to roadside (and vice versa) (A) Unidirectional (B) Bidirectional
(2) Vehicle to vehicle (A) Unidirectional (B) Bi-directional
(3) Selective communication network

In Table 3.4 the type of communication can be identified. This is crucial for the selection of a dedicated type of technology or for the design of a dedicated RTI application using an existing technology. For general driver information a unidirectional communication may suffice, but for toll collection a bidirectional system is necessary. If a proposed system is to be used for a specific application it must fulfil the stated requirements, irrespective of the technical solution. Table 3.5 presents the relevant system performance requirements that are the basis of a choice of a particular product.

Table 3.5   Main parameters describing system performance for
data communication

---

(1) Performance description: message level
   (A) Maximum information bit rate (during a packet) B_{ri}
   (B) Maximum packet repetition time $1/P_r$

(2) Performance description: transmission level
   (A) Maximum range of transmission link
   (B) Maximum vehicle speed
   (C) Maximum bit error probability P_{eo}
   (D) Outage probability P_{out}

(3) Performance description: network level
   (A) Maximum number of users in the network
   (B) Maximum end-to-end transmission packet delay D

Table 3.6 describes the essential technological choices that are able to realise the parameters in Table 3.5, for the different lower OSI layers.

Table 3.6   Main parameters describing system characteristics for data
communication

---

(1) Characteristics at message level
   (A) Maximum transmission bit rate (during a packet) B_r
   (B) Coding technique
   (C) Number of bits per packet n

(2) Characteristics at transmission level
   (A) Carrier frequency f
   (B) Bandwidth B
   (C) Emitted power C_e
   (D) Radiating characteristics (beam width, etc.)
   (E) Minimum signal to noise ratio (at the input of the receiver, to guarantee a required P_{eo}) W_{min}
   (F) Receiver sensitivity
   (G) Equivalent noise power at receiver input (specify the bandwidth) N

(3) Characteristics at network level
   (A) Multiple access protocol
   (B) Switching strategies
   (C) Routing strategies

THE IMPLEMENTATION FILTER

We have now reached the point where the selected application matches certain systems, in both an abstract and a technical manner. The next stage is to determine which systems are most feasible for implementation, taking into consideration factors such as the political climate. We make the assumption here that a sufficient number of systems are viable enough to have reached this stage.

The filter aspects and fulfilment factor

The filter method offers a way of assessing the feasibility of implementing the chosen RTI system. This approach entails appraising each system in terms of several conditions that act as filters to evaluate how easily or successfully a system is likely to be implemented. For each filter described below, the systems are ranked in terms of the extent to which they fulfil that filter's demands. Systems that approximate to an optimal fit to the constraints of any filter are rated highly and assigned high scores, and less effective systems achieve lower scores. Scores from one to ten may conveniently be used. Therefore, that system which best fulfils the filter aspects will receive the value $F = 10$. A system which receives a score of $F = 0$ will be one which is judged to be unrealisable.

There is a risk attached to any estimation of fulfilment. Politicians, parliaments and public authorities may refuse approval of a scheme on grounds other than operational effectiveness, or cost targets may not be met in spite of assumptions made by evaluators. The risk factor $R = 1,...,10$ describes the implications of the non-fulfilment of assumptions underpinning the filter factor F. For each system, the ratio F_a/R_m can be calculated, where F_a = the arithmetic mean of the fulfilment factors for that system, and Rm is the highest value given for any risk factor for that system. The larger the ratio the better the system, that is the best alternative will be indicated by the ratio furthest away from zero.

Filter 1   Comparison with existing systems

The new transmission system under consideration must be evaluated by comparing it with the benefits of existing systems. It should be possible to implement the new system within the framework of existing systems.

Filter 2   Telecommunication regulations

Every country regulates communication services by national telecommunication authorities (in general, the PTTs), and so no communication system may operate outside PTT approval. Therefore when a new transmission system is connected to a public system, it has to conform to existing regulations on a national and international level (or relevant changes in the regulations have to be made by the authorities). The probability that a proposed transmission system will gain such approval is reflected in the F value assigned. For instance, in countries where cable networks are established, it is highly unlikely that the PTT would approve a proposed radio link for communication between fixed entities. With respect to the future mobile communications related to RTI, the PTTs will assess potential systems according to the following criteria:

(1) Telecommunication law: category of application:
   - Individual communication?
   - Distributed communication?
   - Telecontrol problem?

(2) Required transmission capacity:
   - One- or two-way transmission?
   - Frequency bandwidth? Data rate? Data quantity?
   - Range? Attainability?
   - Utilisation frequency, repetition rate?

(3) Suitable postal service:
   - Shared use of GSM?
   - Beacons and (municipal) dedicated lines?
   - Beacons + IDN or ISDN?
   - Beacons with GSM connection to the guidance computer?
   - Shared use of Satcom?

Filter 3   Legal issues

When new transmission systems are installed into an IRTE, they must operate within existing laws. The alternative is for legislative authorities to change existing laws or introduce new ones. If such a law does not exist, consultation with political and traffic authorities may result in the introduction of an appropriate law. Legal issues are based on three main criteria:

(1) Social policy (in the interests of the public):
   - Increased traffic safety?
   - Reduced pollution?
   - Better utilisation of existing roads?

(2) Traffic policy (politically justifiable):
   - Traffic strategy relevant?
   - Benefit/ cost justifiable?
   - Asserted politically?
   - Worthy of subsidisation?

(3) Traffic law (legally flawless):
- Does a legal basis exist?
- Legal responsibility?
- Compatible with road traffic regulations?
- No endangering of traffic safety?
- No data security doubts?

One example of a legal problem relating to RTI systems is the difficulty experienced in using photographs and computer-stored evidence from violation cameras as evidence. In a number of European member states, it has been necessary for police authorities to circumvent the use of such evidence by indirect means. The law needs to be continuously updated to take into account the forms of evidence provided by new RTI systems.

Filter 4   Issues relating to system operation

It is highly likely that systems within an IRTE will be operated by either public transport authorities and/or private organisations. Therefore both options should be accounted for under this filter. The aims will differ considerably, depending upon the private or public status of the organisation, and yet the system should be set up in such a way that their installation is attractive to any operator.

System operators under public law (traffic authorities) would make a detailed review of the benefits of a certain application for, say, reducing congestion; suitability of the technology; whether the operator can assume responsibility for the system; whether public demand and finance justify such a system, etc. The suitability would be judged according to what extent a given system met certain criteria along the lines of the present RTI assessment procedure.

System operators under private law, however, are guided more by economic success. The criteria under this umbrella often overlap those used by traffic authorities, but the "value system" works according to a financial orbit!

As to which sector is preferable, the answers are complex and solutions vary between and within countries. It is not necessarily the case that a country which advocates a "free market", relies solely upon the private sector. For example, France and Italy allow private sector companies to run some motorways.

Filter 5   User issues

A new system has to be attractive to the users, i.e. travellers, otherwise they will not buy it. This aspect relates closely to the one above, as many user issues are settled by the operators. The following aspects are those which the user takes into consideration:

- The expected cost is a very important factor, specifically the procurement and operating costs.
- The quality and reliability should be assured.
- There should be no detrimental effect to the vehicle.
- There should be no disadvantages for the traveller, such as reduced data security or traffic safety.
- Various advantages to the user may increase the attractiveness of the vehicle, such as functions which save time and money; stress-reducing functions; comfort increasing functions, etc.

Filter 6   Aspects of car and system-manufacturers

These aspects are closely related to those mentioned in number 5: a system attractive to the car driver will increase demand, which will in turn increase the production from system manufacturers and the automotive industry.

The risk factor

A risk factor, $R = 1,...,10$ reflects the estimated risk of the non-fulfilment of any decision in the system; this represents the probability that, despite the fulfilment factor estimate F, the system will not meet the demands of the filter. Risks include unrealistic assumptions, governments may refuse approval, or other unforeseen circumstances may alter the basis on which the factor was originally based.

Evaluation

The system to be chosen should be the one with the most favourable ratio of fulfilment factor to risk, that is F_a/R_m. where:

$$F_a = \text{mean of the fulfilment factors of filters}$$
$$R_m = \text{risk estimation 1–10 (maximum)}$$

For example, assume that three alternative RTI systems x, y and z are possible alternative system solutions to the application need (see Table 3.7). The estimated fulfilment factor for each of the six filters outlined above has been calculated. System x is highly compatible with the existing infrastructure (filter 1) but will prove difficult to implement in the current social and/or legal context and looks to have a limited chance of success with users and car manufacturers. This suggests that system x is a technically feasible system that is likely to be socially unacceptable and therefore unlikely to gain favour with legislators. Both systems x and z have high risk factors but y is a low risk alternative.

Table 3.7 Calculation of best alternative system

System	Fulfilment factors for filters 1–6						F_a value	R_m Max	F_a / R_m
	1	2	3	4	5	6			
x	10	7	0	3	2	2	4	6	0.67
y	5	2	4	5	3	1	3.34	2	0.60
z	6	7	8	6	10	9	7.67	6	1.28

To establish which system is the best alternative, the value of F_a (mean of the fulfilment factors of filters), and from it the fulfilment to risk factor relationship, are calculated.

From Table 3.7, system z would be selected as the system with the most favourable F_a / R_m relationship. The final selection of the most appropriate system however, must also take into account the issue of cost.

If and when it has been ascertained that there are several systems that could viably match with the selected application, the probability of implementation can be calculated from an estimation of feasibility and risk. The filter aspects listed here do not represent a definitive array of conditions. Rather, it is a review of those factors considered of key importance.

SUMMARY

This discussion has concerned the process of assessing which systems fulfil particular applications.

Phase one of this process entails gathering a comprehensive list of existing systems (as well as those prototypes in the later stages of development). From this list, the systems which claim to fulfil the chosen application can be identified. In order to match application and system with accuracy, however, the requirements of each can be detailed using a two-stage process. Stage 1 outlines the non-technical specification using the physical model, thereby describing the necessary systems elements. Stage 2 calculates the technical requirements of both the application and system at three levels; message, transmission and network. Figure 3.13 illustrates the decision mechanism at Stage 2.

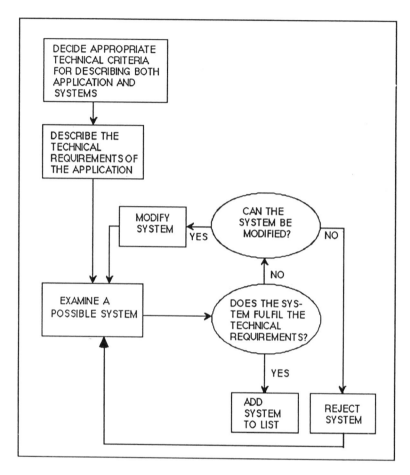

Figure 3.13 Matching systems with the application's technical requirements

Ideally, several systems would meet the application needs, and the con-
cluding decision concerns the selection between satisfactory systems. This
decision makes use of an implementation filter that also deals with non-
technical issues.

4 ASSESSING ALTERNATIVE TRANSMISSION MEDIA

The assessment procedure most commonly adopted would begin with the selection of the application to be fulfilled. Occasions may exist however, in which the procedure begins with an available system for which the most appropriate application must be found. In order to test the suitability of the proposed assessment procedure, a specific RTI application will be considered. In this chapter we shall take as our application the automatic toll debiting on motorways, and apply the assessment procedure up to the point where we can specify the technical requirement. This application is selected partly because it is a well-known case, and its technical content is not innovative. Further discussion of automatic debiting can be found in Hills and Blythe (1990) and in Blythe *et al.* (1991), which give a detailed description of the PAMELA project. Once again, it seems useful to stress that this example is not aimed at describing the application, but at illustrating the way the procedure can be used; therefore the requirements of the system are not described in full detail as this will not compromise the general reliability of the approach. Having summarised the technical requirements of automatic toll debiting we shall consider three systems which might be used to meet these requirement and the requirements of other applications.

TECHNICAL REQUIREMENTS FOR A MOTORWAY AUTOMATIC TOLL DEBITING SYSTEM

An automatic debiting system will allow the driver equipped with the appropriate equipment to carry out toll transactions without stopping at a toll barrier. Our interest is only in the transmission aspects of the air segment, with no consideration of the fixed network devoted to collection of transaction data by the motorway operator. Also, interest is focused on application requirements that are independent both of the final technical solution and the system architecture.

Minimum information bit rate

The information to be exchanged refers to a number of issues, including vehicle identification and classification. Assume that the total amount is equal to 5000 bits. In order to evaluate the maximum information bit rate, we have to know how much time can be used for a transaction. This time interval depends on practical constraints resulting from the minimum separation distance between two consecutive vehicles and from the maximum vehicle speed. Calculation of this requirement will be possible after the selection of maximum vehicle speed and transmission range, considered below.

Maximum packet repetition time

This parameter is appropriate only for applications where the information to be transmitted changes with time and where re-transmission is required.

Minimum range of transmission link

The transaction associated with each operation of automatic debiting can be performed in different ways, but it can be assumed that it will always be carried out in the vicinity of the toll station on a vehicle-by-vehicle basis. Therefore, even though other scenarios might be possible, we consider that the procedure takes place when the vehicle passes through the toll lane. In this case the minimum range of the transmission link is typically a few metres (e.g. 3 m). It is possible to increase the communication link range in order to have a greater time interval for the transaction; however, it should be ensured that just one vehicle per lane is present in the communication zone associated with a single toll beacon. The upper limit of this range will then be equal to the minimum separation distance between two consecutive vehicles. Let us assume a distance of 10 m when vehicles are moving at very low speeds.

Maximum vehicle speed

For the time being, the selection of this parameter is an arbitrary estimation. This value, however, is a key point for calculating many other parameters. Therefore, in order to proceed with this example, a maximum speed equal to 150 km/h will be considered.

Assume that each transaction will be completed before the following vehicle enters the zone illuminated by the toll beacon, (when the speed is equal to 150 km/h the distance is 20 m), a time interval of 480 ms represents the upper limit for the overall procedure to be completed (20 m at 150 km/h). The information bit rate will be then equal to 5000 bits/480 ms (ie. approximately 10.4 kbit/s).

The critical rate is determined by the time the vehicle tag remains in the reading zone. Depending on the transmission technique, the area illuminated by the toll beacon can be much smaller than the separation distance between consecutive vehicles. In such a case the exploitable time interval will be accordingly smaller. Hence a higher information bit rate should be taken into account. However, a larger time interval cannot be achieved, irrespective of technique. Thus the 480 ms interval and the previously computed information bit rate are minimum values which cannot be modified.

Maximum bit error rate probability

The goal of the system operator is usually expressed as a percentage of successful transactions rather than as a bit error rate probability. It is therefore necessary to translate the practical goal of the operator in technical terms. To do this, let us assume the percentage of failed transactions is equal to 0.1% (i.e. a failure will occur for one vehicle per thousand). This percentage will result in an information bit (5000 per transaction) error rate equal to 2×10^{-7}. It is useful to stress that this value is applied to information bits after the decoding process; the error rate of the transmitted bits could be greater.

Outage probability

The probability that the debiting system as a whole fails must be known in operational circumstances. For research systems under test this parameter is usually not defined, as operational and definitive systems will differ often completely from prototypes. The outage probability for the operational system is determined by the operator on the basis of economical considerations (how long can they allow drivers to not pay for road use?). This value is calculated on the basis of an evaluation of a long test period using the system elements of the operational system.

Maximum number of users in the network

Multi-lane architecture results in a number of vehicles participating simultaneously in the payment operation, with the problem of multiple access. Interference can arise from other users accessing adjacent toll beacons. Even with directive beacons some multi-path effects can impair the operation reliability. Single lane transactions occur on a vehicle-by-vehicle basis, so interference from the same lane can be ignored.

The maximum number of simultaneous users, i.e. the number of lanes affecting each other, is set to 5. This choice is justified by considering that in practical situations it is generally sufficient to reject interference coming from the nearest lanes, since the directivity of the beacon prevents interferences originating further away.

Maximum end-to-end transmission delay

This parameter applies when the multiple access scheme selected potentially results in some collisions. The value depends on margins the designer has introduced concerning previous parameters. For instance, if the information bit rate is very close to its minimum value there is no room for possible collisions, because any retransmission would result in an unacceptable information loss. Tables 4.1 and 4.2 summarise the required parameters for application and system features on the three levels of message, transmission, and network.

Table 4.1 Main parameters describing the application requirements of automatic motorway toll collection

(1) Performance description at message level	
(A) Minimum information bit rate	10.4 kbit/s
(B) Maximum packet repetition time	—————
(2) Performance description at transmission level	
(A) Minimum range of transmission link	3 m
(B) Maximum vehicle speed	150 km/h
(C) Maximum bit error probability	2×10^{-7}
(D) Outage probability	not computed
(3) Performance description at 'network level'	
(A) Maximum number of users in the network	5
(B) Maximum end-to-end transmission, packet delay	dependent on 1(A)

A reference to practical solutions has been obtained through the VITA project, which provides the system performance description in Table 4.2. This project was conducted in Spain, France and Italy by motorway operators in conjunction with the Dutch Rijkswaterstaat Rekening Rijden project.

Table 4.2 Main parameters describing system performance for data communication

(1) Performance description at message level	
(A) Minimum information bit rate	11 Mbit/s
(B) Maximum packet repetition time	—————
(2) Performance description at transmission level	
(A) Maximum range of transmission link	7.5 m
(B) Maximum vehicle speed	160 km/h
(C) Maximum bit error probability	10^{-6}
(D) Outage probability	not computed
(3) Performance description at 'network level'	
(A) Maximum number of users in the network	not yet defined
(B) Maximum end-to-end transmission, packet delay	not yet defined

THREE RTI SYSTEMS EXAMINED FOR THEIR SUITABILITY WITH REFERENCE TO AN RTI APPLICATION

Mobile vehicles can only be reached by means of radio. Radio systems can be categorised according to their range of action (see Figure 3.18 in Section

2): *beacons*, which are short -range transmitters (<100 m); *radio concentrators*, which are medium range transmitters (2–20 km); and *satellite radio*, with a range over 20 km. Looking at Figure 3.12 in the previous chapter the link ranges 1 and 2 could use a beacon system, and ranges 3 and 4 could potentially make use of a GSM (Group Special Mobile) infrastructure. However, if different generic systems were used, the problem of standardisation would question its validity. Solutions include the establishment of an IRTE using a modified GSM system for action ranges 1 and 2. The method for evaluating the most appropriate application and system has been described in Chapter 3 (and by von Tomkewitsch and Kossack, 1989, and by Altendorf *et al.*, 1991).

Three systems will now be examined using the assessment procedure. The operation of an automatic debiting system requires communication between the road infrastructure and the road users. Communication could be through a number of alternative transmission media, each requiring their own systems. The three systems to be considered here use roadside cables, microwave sensors, and infra-red. These systems were selected because they were developed within the DRIVE-DACAR consortium, and they illustrate a procedure which can be applied to other systems. For further information about the DACAR project see, for example, Altendorf *et al.* (1991) and van der Hart *et al.* (1991). The following radio systems are still in the development stage, and therefore the technical data remains preliminary and may be incomplete. Because we are starting with a particular system in mind, we will not progress through the assessment procedure in exactly the manner set out in Figure 3.2, but we will use a bottom-up synthesis. This variant of the assessment procedure first requires a description of the system features and elements, and, for systems which can be feasibly implemented, then lists the applications which could be fulfilled (this is the physical model). Application requirements are then listed, in order to determine whether any of the applications match the required system. Each of the matching applications are then costed.

Alternative 1: Cable systems

Cables can be divided into two groups, depending on their location within the infrastructure:

(1) cables in the road (radiating cables)

(2) cables along the road.

Figure 4.1 illustrates this physical distribution. The cables in the road can be responsible for communication between vehicles and roadside infrastructure or, somewhat futuristically, between different vehicles. The cables along the road are responsible for communication between the roadside equipment and supervisory central stations.

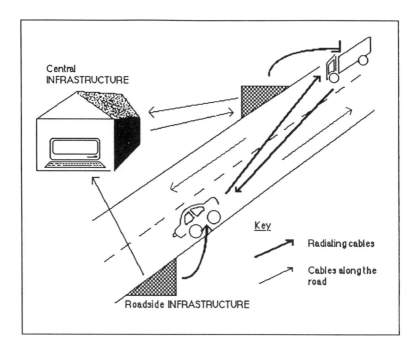

Figure 4.1   Two groups of cables: a schematic diagram

Applications

Radiating cable systems can fulfil the following RTI applications, in addition to automatic debiting for parking or road use:

- general data communications (eg, driver route/parking information);
- automatic vehicle control (eg, automatic vehicle guidance);
- traffic management (eg, control of traffic signals).

System features

The structural description of a cable system is outlined below, prior to an examination of technical features in Tables 4.3 and 4.4. Generally speaking, a continuous cable system will allow vehicles to be guided along a given route according to the cables installed in the lane of the road, i.e. like being driven on an invisible rail. A cable section with a length of about 700 m and an information update rate of once per second enables up to 50 vehicles per cable section to communicate with each other simultaneously. This assumes one vehicle every 14 m and the distance between two vehicles as 10 m (see

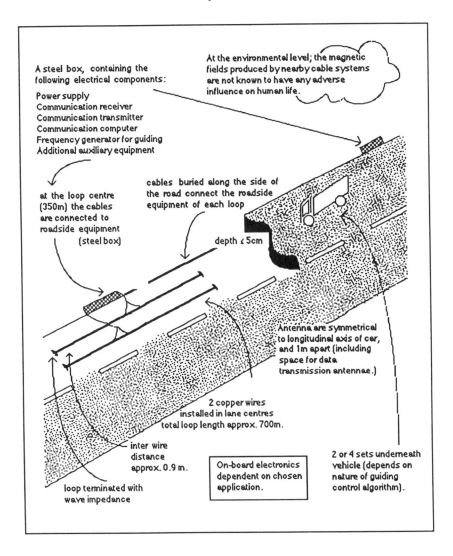

Figure 4.2 The structural description of a continuous cable system

Figure 4.2). Communication between subsequent cable sections can be performed via a roadside computer. The system will operate as a master/slave system with the roadside processor as master. The equipment in the vehicles determines the operational data such as acceleration, velocity and actual longitudinal position and transmits them to the roadside master controller, which in turn determines the required data for optimum driving and sends these to the vehicles. Communication in cable systems is bidirectional; vehicle to vehicle and vehicle to roadside.

Description at transmission system level

Given:	"length of cable section = 700 m	Direct inter-
	"up-date information rate = 1 per second	vehicle
		simultaneous
assuming:	"1 vehicle every 14 m	communication
	"inter-vehicle distance = 10 m	≥ 50 vehicles
Operational data transmitted to "master		Indirect
controller" at the roadside, which then determines		communication
and transmits appropriate data for optimum		(vehicle–
driving to subsequent cable sections, and then to		infrastructure
other vehicles.		–vehicle

The parameters describing the system performance and system characteristics are summarised in Tables 4.3 and 4.4. The following information supplements the data in the tables, and relates to the cable system developed as a demonstration project at the Volkswagen test track (see Ernst, 1991).

- The transmission distance between the vehicle and the cable in the road remains fairly constant during the operation of the system. It depends only on the height of the antennae above the road surface and the position of the vehicle relative to the centre of the lane. This range will be at least 1 m to either side of the centre of the lane. Consequently the distance between the cable and the antenna will not be more than 60 cm.

- The carrier frequency has been set at 113 kHz for the vehicle-to-roadside link and at 69 kHz for the roadside-to-vehicle link.

Table 4.3   Main parameters desc.ribing system performance for cable systems

(1)	Performance description at message level	
	(A) Maximum information bit rate (during a packet) B_{ri}	4.8 kbit/s
	(B) Maximum packet repetition time $1/P_r$	0.5 s
(2)	Performance description at transmission level	
	(A) Maximum range of transmission link	0.6 m
	(B) Maximum vehicle speed	no restriction
	(C) Maximum bit error probability P_{eo}	10^{-7}
	(D) Outage probability P_{out}	10^{-4}
(3)	Performance description at "network level"	
	(A) Maximum number of users in the network	50
	(B) Maximum end-to-end transmission packet delay D	a few ms

- The transmission rate is set at 4.8 kbit/s. It is envisaged to provide 48 packets/s. So a packet contains 100 bits. The effective packet rate is 48 bits per packet. At a repetition rate of 2 there can be 24 packets transmitted at an effective packet rate of 48 bits/packet.

- With an expected intrinsic signal to noise ratio between 15 and 20 dB, a bit error rate (BER) of 10 per million would be sought. The values given are "worst case" values; an actual signal to noise ratio of 20–40 dB is expected. At a noise figure of 5 dB and a receiver band width of 10 kHz, this permits a noise level of − 90 to − 95 dBm/Hz. The expected noise level, however, is − 105 to − 115 dBm/Hz. With a normal distribution of the noise signal, the probability that the allowable noise level is not exceeded is between 99.98 and 99.9997. Only the presence of extraordinary disturbing signals (such as trains with inductive line drives crossing the road), will impair these figures. With the repetition rate of 2, this leads to a packet error rate of 0.001–0.000 01.

Table 4.4 Main parameters describing system characteristics for cable systems

(1)	Characteristics at message level	
	(A) Maximum transmission bit rate (during a packet) B_r	4.8 kbit/s
	(B) Coding technique	see text
	(C) Number of bits per packet n	48
(2)	Characteristics at transmission level	
	(A) Carrier frequency f	69/113 kHz (see text)
	(B) Bandwidth B	
	(C) Emitted power Ce	3 W
	(D) Radiating characteristics (beam, width, etc)	see text
	(E) Minimum signal to noise ratio (at the input of the receiver, to guarantee a required P_{eo}) W_{min}	see text
	(F) Receiver sensitivity	20–40 dB
	(G) Equivalent noise power at receiver input (specify the bandwidth) N	−105 to −115 dBm/Hz
3.	Characteristics "network level"	
	(A) Multiple access protocol	Master/slave
	(B) Switching strategies	
	(C) Routing strategies	see text

- The spectrum efficiency will be 0.24 for this system. However, not all potentially available channels are used.

- The power to be dissipated in the wave resistor closing the cable ends will be 3 W. The term "emitted power" is not applicable, since almost no real power is emitted from the cable.

- The beam configuration parameter is not applicable to cable systems, since the geometry of the magnetic field is pre-determined for a cable in the road and cannot be changed during operation.

- For data communication there are no vehicle speed restrictions. However, if the cable system is used for guiding, there may be limits to the maximum speed, since the curvature of the road and the chosen control algorithm influence the stability of the lateral control of the vehicle.

- As mentioned before, the system can be a master/slave system with the roadside processor as master. With reference to the seven layers of the ISO-OSI model only the physical layer, the data-link layer, and to a certain extent for the continuous cable system the network layer, are relevant to the cable systems. It is clear that the system penetration should be 100% in order to make full use of the advantages of this system. All vehicles have to be guided along this "invisible rail" in order to achieve maximum efficiency in traffic throughput for a given road. With a less complete penetration, the application could be restricted to special users on separate lanes.

- Since the cables are below the surface of the road, it seems difficult to damage them without special tools. The master controllers along the roads have to be installed in solid boxes as are used for telephone distribution stations in urban areas or for electricity distribution substations. The antennae underneath the vehicle are no more vulnerable than other standard car equipment such as mirrors or exhaust pipes. Therefore vulnerability should not be a problem for these components.

We will now examine the following cable systems:

- continuous cable system;
- discrete cable system;
- passive inductive beacon.

Their characteristics will be determined to see how each system matches the technical requirements of the application. For this example the applications have been categorised according to Table 3.1 (see Chapter 3).

Matching application with system

The three cable systems mentioned above (continuous cable systems, discrete cable systems and passive inductive beacons), have been checked for their suitability for the various applications. Table 4.5 presents a list of applications, where the (total or partial) implementation of each type of cable system has been indicated by crosses.

On the level of systems application a distinction has been made of the degree of fulfilment each system may achieve. The following symbols are used to denote degrees of fulfilment:

x   means fits the application
(x) means that the system may fit the application
(*) means that the system may be part of a total system for the application concerned.

Table 4.5 summarises the applications where cable systems are considered appropriate, but we also need to differentiate between types of cable system because they have different characteristics and suitability to different applications.

Continuous cable system

Continuous cable systems require 100% system penetration in order to derive full benefits; therefore it is unlikely that they will be rapidly implemented on a large scale. However, when there is heavy traffic demand, such as limited available space in the lateral direction (e.g. special highways through tunnels or over bridges), the introduction of a continuous cable system may be considered. For other applications, vehicles of the future may be equipped with "drive by wire" systems for steering, braking and accelerating (cf. the Volkswagen Tempomat, ABS). The antennae required for matching these cars to a cable track pose only a minimal extra cost. For these reasons continuous cable systems have received an "*x*" for application 1.5 (automatically guided systems) and for application 8.2 (public transport guidance systems).

In principle, a continuous cable system can meet the same demands as the applications of a discrete cable system, though probably not at the same time due to the limited transmission rate. Great advances have been made, however, in the development of data compression techniques, although their consequences for RTI have not been investigated. As continuous cable systems are limited to special lanes or tracks, it remains of limited use for a complete IRTE. For this reason, continuous cable systems have not been marked for the other applications.

Discrete cable system

Both discrete cable systems and passive inductive beacons are types of inductive loop. Before describing these two cable systems, it is first necessary to introduce inductive loops.

An inductive loop is analogous to a bar magnet in the way that it operates. An alternating current passes through a loop, generating a fluctuating magnetic field around the wires. When a metal object enters the field it induces eddy currents from its own magnetic field, which decrease the loop's inductance by a quantifiable amount. Therefore, when a vehicle enters the zone of detection, an inductance change occurs which is detected by the loop detector circuitry. Continuous tracking of the inductance prevents random fluctuations due to environmental conditions being detected as vehicles.

A discrete cable system consists of an inductive loop in the road. It can be used for incident detection and for communication. The transmission rate of this system for communication will depend on the size of the loop, required transmission quality and vehicle speed. From the small amount of experimental investigation, ways in which this system can be integrated into an IRTE are outlined below.

A discrete cable system can be part of a total RTI system. This can be seen from Table 4.5, where a majority of the applications received a "(*)". This criterion was developed by BMC in the DACAR consortium. A detailed evaluation in terms of the criteria, both of the potential advantages and limitations of this system, is required before further judgement can be made. Such evaluation must be based upon empirical assessment.

For most traffic control functions, discrete cable systems are a valuable tool. This applies particularly to incident detection (applications 3.1, 3.2 and 6.1 to 6.4) and data transmission (applications 2.1 to 2.4, 4.1 to 4.4 and 5.1 to 5.3). Discrete cable systems can also be used for commercial and public fleet management (application 8.1) and for automatic debiting systems (applications 9.1 to 9.4). For the complete list of applications, see Table 4.5.

Inductive loops are relatively easy to install, and do not require extensive roadside excavation. The operational lifetime is long if the system is correctly installed, and advances in electronics have increased the robustness during operation. In addition, they are not generally vulnerable to damage by vehicles.

Passive inductive beacons

This is another inductive loop system, operating independently of external power sources. It receives its operational energy from the vehicle approaching the loop. Insufficient information is available on such a device to fully evaluate this system according to the criteria. It seems that such a device will have a very limited transmission rate (as have the other inductive systems), but its advantages are that it may be easily implemented and has low costs. As can be seen from Table 4.5, a passive inductive beacon may be part of a total system for applications 2.1 and 5.1.

Table 4.5 Applications for cable systems

Function group	Functional area		Applications	Cable system Continuous	Cable system Discrete	Passive inductive beacon
Traffic control systems / Driving support systems	Cooperative driving systems	1.1	Intelligent manoeuvring systems			
		1.2	Medium-range preinformation systems		(*)	
		1.3	Data collection for traffic management		(*)	
		1.4	Intersection control systems		(*)	
		1.5	Automatically guided systems	x		
	Route guidance systems	2.1	Actualisation of autonomous route guidance systems		x	
		2.2	Dynamic route guidance, traffic dependent, locally		(x)	
		2.2	Dynamic route guidance, traffic dependent, regionally		(*)	
		2.4	Long-distance route guidance during holidays		(*)	
	Parking information systems	3.1	Operation of parking gates and garage doors		x	
		3.2	Parking management, park and ride systems		(x)	
		3.3	Booking of parking spaces for passenger cars			
		3.4	Booking of parking spaces for trucks (transit traffic)			
		3.5	Booking of reserved space for just in time delivery			
	Driver info systems	4.1	Local DIS on road works, detours etc.		(*)	
		4.2	Regional DIS on road works, detours etc.		(*)	
		4.3	Local non-traffic-oriented DIS		(*)	
		4.4	Regional non-traffic-oriented DIS		(*)	
	Warning system	5.1	Transmission of warning signals into the car		x	(*)
		5.2	Warning of accidents and incidents		(*)	
		5.3	Emergency call systems		(*)	
	Traffic mgt info system	6.1	Optimisation of traffic lights at single intersections		x	
		6.2	Local coordination of traffic lights		x(*)	
		6.3	Traffic management in cities		(*)	
		6.4	Regional traffic management		(*)	
		7	Tunnel management systems			
Special services	Commercial & public fleet mgt	8.1	Transmission of data from and into vehicles		(x)	
		8.2	Public transport guidance systems	x		
		8.3	Regional fleet management (taxis, police)			
		8.4	Long-distance fleet mgt (truck fleets, just in time del.)			
	Automatic debiting systems	9.1	Automatic fee collection systems on freeways		(x*)	
		9.2	Automatic fee collection systems of parking lots		(x*)	
		9.3	Road pricing		(x*)	
		9.4	Vehicle taxation on the territorial principle		(x*)	

x = system fits application
(x) = system may fit (but not perfectly)
(*) = system may fit into IRTE application

Alternative 2: Microwave systems

Microwave systems can be classed as vehicle sensors which offer an alternative approach to loops, which are described above, and tubes. These systems work on the Doppler principle, whereby an energy beam is directed at the road, and reflected back by moving objects. Unfortunately it does not detect stationary objects and small vehicles are often missed in the shadow of larger ones. There are suggestions that the first problem could be overcome, but the second drawback may well remain.

Technological progress is leading to the development of cost effective devices operating on additional band frequencies; therefore new bands will have to be made available. Also the higher the frequency, the smaller the equipment size, which reduces the problem of integrating the equipment into the car. Future use of frequency bands, which are not at present allocated, will ease the problems relating to pre-existing users sharing the same spectrum resources. At the beginning of 1991 the CEPT proposed that the 64–65 GHz band be used for future RTI systems (see also Boheim and Fischer, 1991).

The present description relates to a demonstration system developed by Marconi Command and Control Systems Limited and Standard Elektrik Lorenz AG suitable for a large number of RTI applications. It was demonstrated on normal roads in 1991 in Germany and the UK (see Höfgen *et al.* 1991).

Marconi/SEL have designed a link to operate at a transmission frequency of 60 GHz. The choice of this frequency band allows advantage to be taken of the peak derived from atmospheric oxygen attenuation at this frequency. This means lower mutual interference among adjacent roadside units, which allows the use of the same operating channel and frequency.

There is a wide range of potential applications for microwave systems. Provided the information to be transmitted can be imposed on the available bandwidth, all short-range point-to-point communications are feasible. We will restrict this assessment of microwave systems to one in particular: the Marconi/SEL Microwave System.

Applications

Some of the applications for a roadside-to-vehicle link so far considered are listed below (all require a bandwidth of less than 1 MHz):

- route guidance
- parking
- driver information

- emergency warnings
- fleet management
- tunnel management
- automatic debiting.

The vehicle-to-vehicle link will operate as a support to safe driving, carrying the following information:

- vehicle speed
- acceleration/deceleration
- driver's intent.

Future developments may include anti-collision capabilities, but this would require a much greater bandwidth.

System features

We will begin with a structural description of the microwave system, from the transmission to the roadside level. Then the technical features will be examined in Tables 4.6 and 4.7.

On the transmission level, microwave systems can sustain all operation modes associated with RTI systems, but are not suitable for broadcasting services (see Table 4.8).

The vehicle-mounted transceiver is small, making it easy to install in the body of the car. Antennae should be located externally in order to reduce propagation losses. They can be positioned behind light clusters, behind front or rear number plates, inside plastic wing mirror housings, at the top of the windscreen or the rear window. Some prototypes are being constructed using waveguide components. In order to avoid the development of new interfaces, a waveguide horn radiator is used. Operational apparatus is equipped with miniaturised devices in a low-cost and integrated assembly.

The dielectric characteristics of the window protecting the antennae need to be taken into account, although their effect on expected performance may be negligible. The vehicle-mounted unit consists of a millimetre wave/intermediate frequency subsystem which is connected to a processing unit which recovers the base-band signal. Particular attention must be paid to minimising power consumption, especially when the vehicle is stationary.

Roadside devices are expected to be spaced at about 1 km with an operational range of about 300 m. In order to cope with unfavourable weather conditions, variable transmitting power and de-icing devices will be provided.

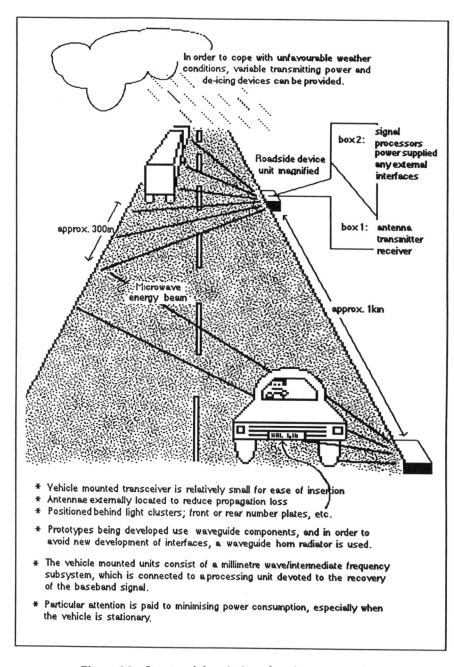

In order to cope with unfavourable weather conditions, variable transmitting power and de-icing devices can be provided.

box 2: signal processors power supplied any external interfaces

Roadside device unit magnified

box 1: antenna transmitter receiver

approx. 300m

Microwave energy beam

approx. 1km

HAL LIB

* Vehicle mounted transceiver is relatively small for ease of insertion
* Antennae externally located to reduce propagation loss
* Positioned behind light clusters; front or rear number plates, etc.

* Prototypes being developed use waveguide components, and in order to avoid new development of interfaces, a waveguide horn radiator is used.

* The vehicle mounted units consist of a millimetre wave/intermediate frequency subsystem, which is connected to a processing unit devoted to the recovery of the baseband signal.

* Particular attention is paid to minimising power consumption, especially when the vehicle is stationary.

Figure 4.3　Structural description of a microwave system

Table 4.6 Main parameters describing system performance for a microwave link

(1) Performance description at message level
    (A) Maximum information bit rate (during a packet) B_{ri} 274 kbit/s
    (B) Maximum packet repetition time $1/P_r$ 0.1 s   (1)

(2) Performance description at transmission level
    (A) Maximum range of transmission link   300 m
    (B) Maximum vehicle speed   200 km/h
    (C) Maximum bit error probability P_{eo} 5.10×10^{-3} (*)
    (D) Outage probability $<4.10 \times 10^{-3}$ (2)

(3) Performance description at "network level"
    (A) Maximum number of users in the network   30   (3)
    (B) Maximum end-to-end transmission packet delay D 0.1(n–1) s   (4)

(*) = without error correction coding
(1) Assume transmission bandwidth of 500 kHz. Potentially the bandwidth could be increased to 10 MHz.
(2) Probability of not being able to use any given 150 bit message due to transmission errors. Outage probability due to atmospheric effects is 1.10 x 10-3.
(3) Many more vehicles could receive a signal in a beacon–vehicle communication.
(4) Where n is the number of vehicles in any given chain. There is no theoretical limit to n, although in practice, it is unlikely that n will be greater than 20, i.e. 20 vehicles each separated by 50 m on a motorway.

Table 4.7 Main parameters describing system characteristics for a microwave link

1. CHARACTERISTICS AT MESSAGE LEVEL
    (i) Maximum transmission bit rate (during a packet) B_r 500 kbit/s   (5)
    (ii) Coding technique   Reed-Solomon   (6)
    (iii) Number of bits per packet n   150   (7)

2. CHARACTERISTICS AT TRANSMISSION LEVEL
    (i) Carrier frequency f   60 GHz
    (ii) Bandwidth B   500 kHz   (8)
    (iii) Emitted power C_e 20 mW
    (iv) Radiating characteristics (beam, width, etc) 20^0 to 3dB points
    (v) Minimum signal to noise ratio (at the input of
         the receiver, to guarantee a required P_{eo}) W_{min} 20 dB
    (vi) Receiver sensitivity 10×10^{-13} W   (9)
    (vii) Equivalent noise power at receiver input
         (specify the bandwidth) N 2.5×10^{-14} W   (9)

3. CHARACTERISTICS AT 'NETWORK LEVEL'
    (i) Multiple access protocol   CSMA/CD   (10)
    (ii) Switching strategies   CSMA/CD   (10)
    (iii) Routing strategies   (11)

Table 4.7 Notes (*continued*)

(5) Potentially bandwidth could be increased to 10 MHz, limited only by the coherence bandwidth. Resulting bit rate: 10 Mbit/s.

(6) Other coding techniques also possible.

(7) Assumes 548 ms packet.

(8) Potentially bandwidth could be increased to 10 MHz, limited only by coherence bandwidth.

(9) Assuming 500 kHz bandwidth.

(10) Carrier sense multi-O access with collision detection. Other protocols and switching strategies also possible.

(11) Routing defined at time of use by access protocol.

Units should be contained in two boxes: antenna, transmitter and receiver in one box and signal processors, power supplies and any external interfaces in the second box.

Both bidirectional and unidirectional types of communication are feasible for vehicle-to-roadside and vehicle-to-vehicle communication, as well as a selective communication network.

Figure 4.3 illustrates this structural description of the vehicle and roadside levels.

Matching application with system

Table 4.8 examines which system elements match which applications.

Alternative 3: Infra-red systems

Infra-red systems are another alternative to the use of cables. The system operates by electromagnetic waves. Their wavelength is somewhat longer than the wavelength of visible light. The present discussion makes particular reference to the VALEO infra-red system for vertical transmission. However, some of the discussion will also be relevant to systems such as Autoguide and LISB which transmit in a horizontal plane. The VALEO infra-red transmissions concept has been demonstrated at the Mortefontaine test track, near Paris, in 1990–91 (see van der Hart *et al.* 1991)

Electromagnetic waves are subject to free space loss, absorption, reflection, interference etc., and limitations like the optical horizon. Absorption can be kept at a feasible order of magnitude by the choice of the wavelength

Table 4.8 Applications for microwave systems

Function group	Functional area	Applications	Link mode: vehicle to vehicle	Link mode: vehicle to beacon
Driving support systems	Cooperative driving systems	1.1 Intelligent manoeuvring systems	◆	
		1.2 Medium-range preinformation systems		x
		1.3 Data collection for traffic management		x
		1.4 Intersection control systems		x
		1.5 Automatically guided systems	◆	
Traffic control systems	Route guidance systems	2.1 Actualisation of autonomous route guidance systems		x
		2.2 Dynamic route guidance, traffic depentent, locally		x
		2.2 Dynamic route guidance, traffic depentent, regionally		x
		2.4 Long-distance route guidance during holidays		x
	Parking information systems	3.1 Operation of parking gates and garage doors		x
		3.2 Parking management, park and ride systems		x
		3.3 Booking of parking spaces for passenger cars		x
		3.4 Booking of parking spaces for trucks (transit traffic)		x
		3.5 Booking of reserved space for just in time delivery		x
	Driver info systems	4.1 Local DIS on road works, detours etc.		x
		4.2 Regional DIS on road works, detours etc.		x
		4.3 Local non-traffic-oriented DIS		x
		4.4 Regional non-traffic-oriented DIS		x
	Warning system	5.1 Transmission of warning signals into the car		x
		5.2 Warning of accidents and incidents		x
		5.3 Emergency call systems		x
	Traffic mgt info system	6.1 Optimisation of traffic lights at single intersections		x
		6.2 Local coordination of traffic lights		x
		6.3 Traffic management in cities		x
		6.4 Regional traffic management		x
		7 Tunnel management systems		x
Special services	Commercial & public fleet mgt	8.1 Transmission of data from and into vehicles		x
		8.2 Public transport guidance systems		x
		8.3 Regional fleet management (taxis, police)		x
		8.4 Long-distance fleet mgt (truck fleets, just in time del.)		x
	Automatic debiting systems	9.1 Automatic fee collection systems on freeways		x
		9.2 Automatic fee collection systems of parking lots		x
		9.3 Road pricing		x
		9.4 Vehicle taxation on the territorial principle		x

◆ = improved by inclusion of anti-collision secondary radar

x = system fits application

(x) = system may fit (but not perfectly)

(*) = system may fit into IRTE application

at 850 nm. Choice of wavelength is also determined by the technological possibility of constructing transmitters and receivers for this wavelength. Transmission and reception gains can be obtained by optical focusing devices such as lenses. Infra-red waves can be modulated by all known modulation techniques and thus can carry information. As the wavelength mentioned (850 nm) represents an extremely high frequency (in comparison to radio waves and microwaves) the bandwidth available for information transfer is also large. An infra-red communication link can be expressed in a link budget. The free-space loss and absorption limits the distance. At the same time the multiple path effect is minimal. No significant provision for fading needs to be made. Finally, infra-red devices have low energy needs and low production costs.

Applications

Available products based on the use of infra-red radiation for carrying information are:

- remote control of electronic equipment for entertainment purposes (radio, TV, video recorders, CD players etc.);
- remote control for locking and unlocking doors of vehicles;
- automobile route guidance systems , installed in London and Berlin;
- communication links of the car-to-beacon type.

Potential applications of infra-red systems are listed in Table 4.11.

System features

The particular infra-red system to be explored in this section is the VALEO experimental programme within the DACAR project. Two prototype applications have been developed within VALEO:

- a mileage beacon
- a guidance beacon.

The usefulness of the message capacity for activating a smart signalling panel was investigated by VALEO and Supelec. This subsection will outline the structural description, then the technical features of the mileage beacon system will be examined in Tables 4.9 and 4.10.

Mileage reference beacons

A mileage beacon can emit data concerning the position of the vehicle. It provides information on the type of road, the road number and the position along the road of the reference beacon, thus allowing drivers to obtain data relating to their position. Experiments have been carried out with beacons on gantries at a height of >5 m, which is suitable for nearly all road situations. On each gantry there is a beacon for every lane. As the system provides mileage reference information, the gantry mounted beacons all send the same information.

When transmitting from a height of 5 m above the road, the beam at ground level can have a width of 2.5 m and a thickness of 14 cm. The dimensions given are simply representative of the experiments. Other dimensions of the beam can be created as required.

With these dimensions it has been proved that a message of 50 bits can be received reliably under all weather conditions, and on multiple lane roads. The ratio between the message and the inter-message gap is of the order of 100. This means that many more messages can be conveyed, providing that the receiving equipment is capable of selecting the information desired by the driver at that particular time. The distance between beacons for a mileage reference application can be tens of kilometres. Once a vehicle has received the information regarding its location from a beacon, an on-board computer which records time and vehicle speed can update the information for a long distance and so keep in touch with real traffic conditions without the need for frequent updates from a beacon. However, if additional information is also conveyed to the driver, the distance between gantries may need to be decreased. The only type of communication possible is roadside to vehicle (unidirectional).

Table 4.9 Main parameters describing system performance for an infra-red mileage beacon

(1) Performance description at message level	
(A) Maximum information bit rate (during a packet) B_{ri}	493 kbit/s
(B) Maximum packet repetition time $1/P_r$	continuous
(2) Performance description at transmission level	
(A) Maximum range of transmission link	6 m+
(B) Maximum vehicle speed	120 km/h
(C) Maximum bit error probability P_{eo}	0.01 (in 4 messages)
(D) Outage probability P_{out}	
(3) Performance description at "network level"	
(A) Maximum number of users in the network	
(B) Maximum end-to-end transmission packet delay D	

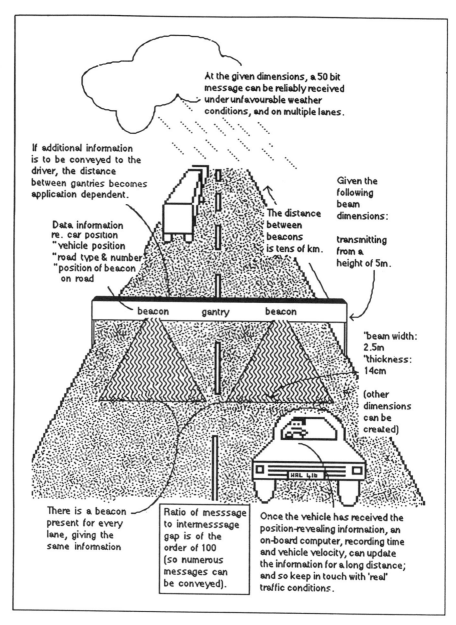

At the given dimensions, a 50 bit message can be reliably received under unfavourable weather conditions, and on multiple lanes.

If additional information is to be conveyed to the driver, the distance between gantries becomes application dependent.

Data information re. car position
"vehicle position
"road type & number
"position of beacon on road

The distance between beacons is tens of km.

Given the following beam dimensions:

transmitting from a height of 5m.

beacon   gantry   beacon

'beam width: 2.5m
'thickness: 14cm

(other dimensions can be created)

There is a beacon present for every lane, giving the same information

Ratio of messsage to intermesssage gap is of the order of 100 (so numerous messages can be conveyed).

Once the vehicle has received the position-revealing information, an on-board computer, recording time and vehicle velocity, can update the information for a long distance; and so keep in touch with 'real' traffic conditions.

HRL 4lb

Figure 4.4   Structural description of a mileage reference beacon

Table 4.10 Main parameters describing system characteristics for an infra-red mileage beacon

---

(1) Characteristics at message level
- (A) Maximum transmission bit rate (during a packet) B_r 650 Kbit/s
- (B) Coding technique   PSK 2
- (C) Number of bits per packet n 51

(2) Characteristics at transmission level
- (A) Carrier frequency f 850 nm
- (B) Bandwidth B
- (C) Emitted power C_e
- (D) Radiating characteristics (beam, width, etc.)
- (E) Minimum signal-to-noise ratio (at the input of the receiver, to guarantee a required P_{eo}) W_{min} 9 dB
- (F) Receiver sensitivity
- (G) Equivalent noise power at receiver input (specify the bandwidth) N

(3) Characteristics at "network level"
- (A) Multiple access protocol
- (B) Switching strategies
- (C) Routing strategies

Other beacons

Beacons have several different uses and vary in the medium they use for transmitting information. Beacons used for route guidance will be located in a different way than those used for road pricing. Similarly, if the beacons are to be used in an urban environment then a larger number will be needed than if they are to be used in an inter-urban area.

The technical restrictions of the different transmitting media will also affect decisions about how many beacons are necessary. Microwave transmitters have a much greater capacity for transmitting information than, for example, a so-called "leaky feeder" cable.

The question of how many beacons to use is essentially an empirical issue. It is also an area in which there has only been a limited amount of research. Experimental systems in London have used less than ten beacons and although experience in highway use is increasing, it is probably important to interpret figures for the numbers of beacons used with care.

Beacon systems other than an infra-red mileage beacon include:

- A broadcasting beacon with an omnidirectional beam, for communication of longer messages; consisting of an alpha numerical text of about 122 characters (976 bits effective; nominal 1019 bits);

Table 4.11   Applications for infra-red systems

Function group	Functional area		Applications	mileage beacon	guidance beacon
Driving support systems	Cooperative driving systems	1.1	Intelligent manoeuvring systems		
		1.2	Medium-range preinformation systems		
		1.3	Data collection for traffic management		
		1.4	Intersection control systems		x
		1.5	Automatically guided systems		
Traffic control systems	Route guidance systems	2.1	Actualisation of autonomous route guidance systems	x	x
		2.2	Dynamic route guidance, traffic depentent, locally		x
		2.2	Dynamic route guidance, traffic depentent, regionally		x
		2.4	Long-distance route guidance during holidays		x
	Parking information systems	3.1	Operation of parking gates and garage doors		x
		3.2	Parking management, park and ride systems		
		3.3	Booking of parking spaces for passenger cars		(*)
		3.4	Booking of parking spaces for trucks (transit traffic)		(*)
		3.5	Booking of reserved space for just in time delivery		(*)
	Driver info systems	4.1	Local DIS on road works, detours etc.		x
		4.2	Regional DIS on road works, detours etc.		x
		4.3	Local non-traffic-oriented DIS		x
		4.4	Regional non-traffic-oriented DIS		x
	Warning system	5.1	Transmission of warning signals into the car	x	x
		5.2	Warning of accidents and incidents	(*)	x
		5.3	Emergency call systems		x
	Traffic mgt info system	6.1	Optimisation of traffic lights at single intersections		(*)
		6.2	Local coordination of traffic lights	(*)	(*)
		6.3	Traffic management in cities	(*)	(*)
		6.4	Regional traffic management		
		7	Tunnel management systems		
Special services	Commercial & public fleet mgt	8.1	Transmission of data from and into vehicles	(*)	
		8.2	Public transport guidance systems		(*)
		8.3	Regional fleet management (taxis, police)		(*)
		8.4	Long-distance fleet mgt (truck fleets, just in time del.)		
	Automatic debiting systems	9.1	Automatic fee collection systems on freeways		(*)
		9.2	Automatic fee collection systems of parking lots		(*)
		9.3	Road pricing		(*)
		9.4	Vehicle taxation on the territorial principle		(*)

x = system fits application
(x) = system may fit (but not perfectly)
(*) = system may fit into IRTE application

- A beacon with a wide beamwidth for communication with a vehicle during a relatively long period of passing through the beam, which enables reception of very long messages (up to 17 400 bits for a vehicle with a speed of 180 km/h).

Vertically polarised infra-red transmissions are effective over relatively long distances. The disturbance by sunlight is noticeable but small, and because of the vertical transmission, the influence of rain is low.

Matching application with system

Table 4.11 indicates which system elements match with which (potential) RTI applications. The crosses and asterisks indicate for which of the applications infra-red transmissions are appropriate. This does not, however, mean that infra-red technology is unsuitable for other applications.

SUMMARY

This chapter supplements the assessment procedure by implementing dissected parts of the procedure using the relevant technical information from specified examples. The first part of the chapter takes the application *Motorway Automatic Toll Debiting System*, but does not implement the procedure beyond the point of specifying the technical requirements of the application. The second part examines three systems; cable, microwave and infra-red. The assessment procedure thus begins with the system. After a brief outline of the potential applications for each system, the system features and technical requirements are stated prior to a table showing which systems match with which applications.

Section 3
TECHNICAL DEVELOPMENTS

5 DYNAMIC ROUTE GUIDANCE: SEARCH TECHNIQUES

This section of provides a technical background to support the assessment procedure developed in Section 2.

A principal application of RTI, in the sense of visibility to the road user, is route guidance, and products are already being marketed to provide some information to drivers in transit. We have already mentioned systems such as the ALI, ALI-SCOUT, Autoguide and EURO-SCOUT "family" which provide up-to-date route advice, initially on the German autobahn system, and now more widely. Catling *et al.* (1991) present an overview of an attempt to use cellular radio as the basis for an integrated road transport environment in which dynamic route guidance (DRG) is the primary application. French and Lang (1973) provided one of the earliest descriptions of automatic route control, and the ALI-SCOUT and Autoguide systems have been on trial for some time now. See, for example, von Tomkewitsch (1987), Bonsall *et al.* (1991), and Bright and Ayland (1991) on the ALI-SCOUT trials in Berlin (the LISB project), and Catling (1987), Jeffrey (1987), and Jeffrey *et al.* (1987) on the Autoguide trials in London. We shall also present an evaluation of alternative route guidance systems for the city of Amsterdam, using a cost–benefit analysis, in Chapter 7.

Further discussion of the current development of DRG systems can be found in a series of papers from the CARGOES consortium and published in the DRIVE Conference Proceedings by Hounsell *et al.* (1991), by Bolelli *et al.* (1991a, b), by Charbonnier *et al.* (1991), by Aicher *et al.* (1991), and by Beccaria *et al.* (1991). These papers consider the problems associated with modelling road usage and data management in an integrated dynamic route guidance and traffic control system .

Dynamic route guidance is a popular application of RTI partly because of the marketability of the product to the individual motorist, and we can expect to see rapid growth in these commercial navigational aid systems. A fundamental problem to be solved by any DRG system which has information about the vehicle's present and intended locations, about available routes, and about congestion, is to decide upon the optimal route. This decision about the route between the present and intended locations can be regarded as a real-life example of the "travelling salesman problem" much beloved by researchers in artificial intelligence (AI). Route choice calculations in DRG are made using algorithms developed in AI laboratories to solve problems by exploring the available search space—algorithms designed to find the shortest route through a well-defined network.

In this chapter we shall focus upon the choice of the optimal route, but it is first necessary to review the system components. A fully dynamic route guidance system comprises a number of system elements and functions. These can be categorised as follows:

- user interfaces

- microprocessors

- memory requirements

- dynamic input data

- route choice calculations.

User interfaces need to provide the driver with an easy-to-use system, presenting information in a clear, precise format. Current systems allow the user to enter an origin and destination, and in some cases a route selection criterion. Those systems providing a map display in the vehicle offer alternative map scales and levels of detail. A number of system manufacturers are expending significant effort on determining the human–computer interface characteristics of the in-vehicle equipment. There is an obvious trade-off between providing the user with all the information required, and ensuring that the driver is not distracted from the driving task.

The requirements of the microprocessor vary according to the functions offered by the system. For a dynamic route guidance system, the response times for each function depend on the microprocessor power. The functions controlled by the microprocessor include accessing the data base, processing dynamic input data, performing dead-reckoning and map-matching functions, redrawing in-vehicle maps based on a changed vehicle position or scale requirement, and calculating the optimum route.

Memory requirements for autonomous or externally linked route guidance systems are potentially very large. An autonomous system needs sufficient memory to hold a highway network data base down to a resolution appropriate for route choice calculations and map displays. Urban and national networks can be stored in removable modules of various kinds to reduce the size and cost of the in-vehicle unit. The second memory requirement is to hold the software which gives the system its functionality and can implement the necessary calculations for route guidance.

The dynamic input data communicated to an in-vehicle route guidance system typically provides information on actual journey times and delays. This can serve to update historical information held in memory. A dynamic route guidance system could make use of data in various formats (such as the type proposed for RDS-TMC). The data are interpreted to allow optimum route selection in real time, thus resulting in a system that responds to actual traffic conditions.

The specific function of a dynamic route guidance system of particular interest here concerns the route choice calculations. This is one of several functions performed by the microprocessor, based on the highway network data base and the dynamic input data. The calculations are performed in-vehicle by some systems, or at a central computer by others. The algorithm used to perform these calculations is paramount to the success of a dynamic system, particularly where calculations are performed in-vehicle. Ideally, the algorithm needs to be sufficiently detailed to allow all relevant routes to be considered, but simple enough to give a response within a short time period.

There are a number of algorithms which lend themselves to the calculation of optimum routes in a highway network. Several specific algorithms used by current systems are described below.

APPROACHES TO ROUTE CHOICE CALCULATION

An algorithm can be defined as a mechanical or recursive computational procedure which uses a set of rules to perform calculations. There are a number of algorithms available to find the shortest path through a network. Many of these depend on search methods, such as those described by Nilsson (1980) and Pearl (1984). Three of the basic search methods are:

- breadth first
- depth first
- best first.

Breadth first search

Any route choice problem can be represented by a series of nodes and links. An example of such a problem state is a driver wishing to travel from A (origin) to B (destination). As a first step, there is the option of travelling from A to either C or D (Figure 5.1). If the driver selects the route to C, he can then travel to E, F or G. If D is selected, he can travel to H, I or J (Figure 5.2). The problem space is expanded until an acceptable solution emerges (Figure 5.3). The search strategy determines the chosen method of expanding the problem space.

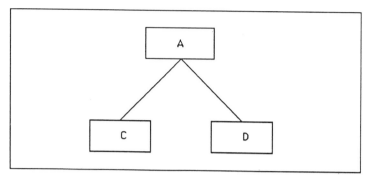

Figure 5.1   Breadth first search, step 1

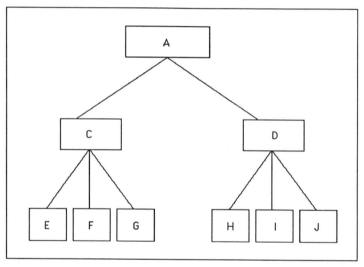

Figure 5.2   Breadth first search, expanded problem space

The breadth first search method expands all nodes at each step. This approach is guaranteed to find a solution if one exists. The level of detail involved in the breadth first search ensures that all possible solutions are considered, thus ensuring that the best solution will be found.

There are a number of drawbacks with the breadth first search method, however. Firstly, the memory requirements whilst performing the calculations are considerable. As each node is expanded fully at each level, the memory requirements increase exponentially as the levels increase. The second drawback of this method is the calculation time. Again, as each node is expanded, the number of calculations required increases exponentially with network size. Finally, for a large network, the number of potential routes which need to be examined causes the method to become unmanageable, thus rendering the breadth first search method inappropriate.

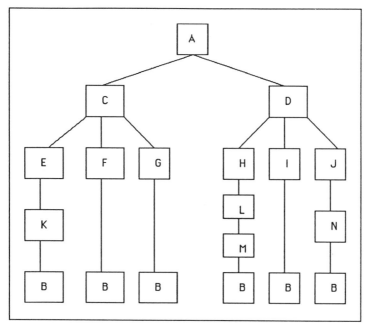

Figure 5.3 Breadth first search solutions

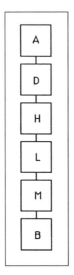

Figure 5.4 Depth first search

Depth first search

An alternative way of expanding the problem uses the depth first search method. This method expands one node at each step. The second node is then expanded in the same manner. This process is repeated until the destination is reached (Figure 5.4).

The depth first search method has the advantage of requiring considerably less memory than the breadth first search. By examining fewer nodes, the calculation time is significantly shorter. However, as shown in Figure 5.4, although the depth first search method is quicker than the breadth first search, the route selected may not necessarily be a direct route.

The depth first search method probably represents the approach used by most drivers in selecting a route based on local knowledge. It has no practical application in computerised route guidance systems, except to illustrate an extreme.

Best first search

The cases outlined above represent the two extremes in search methods. Best first search is a way of combining the advantages of both of these methods. In the best first search method, the first node is fully expanded. The most promising node expanded from the first node is then selected for further expansion. A suitable heuristic function needs to be applied in order to determine the most promising node.

In the best first search method, the second node is then expanded. If this produces a solution, the process is complete. Otherwise, all new nodes are added to the set of nodes which have so far been generated but not expanded. From this set, the most promising node is again selected and the process is repeated.

In normal operation, there is a period of depth first search as a promising route is explored. Eventually, if a solution is not found, the route becomes less promising than one of the top level routes which has previously been ignored. The more promising route is then expanded until a solution is found or it becomes less promising. This process is illustrated in Figure 5.5 and described below:

Step 1   The origin is selected as node A.

Step 2   Node A is expanded, yielding the set of nodes B, C and D. The heuristic values of these nodes are 3, 5 and 1, respectively.

Step 3   Node D, having the lowest heuristic value, is selected for further expansion. This yields nodes E and F with heuristic values 4 and 6, respectively.

Step 4   Node B then has the lowest heuristic value and is expanded. This yields nodes G and H with heuristic values 6 and 5, respectively.

Step 5   Node E then has the lowest heuristic value and is expanded. This yields nodes I and J with heuristic values 2 and 1, respectively.

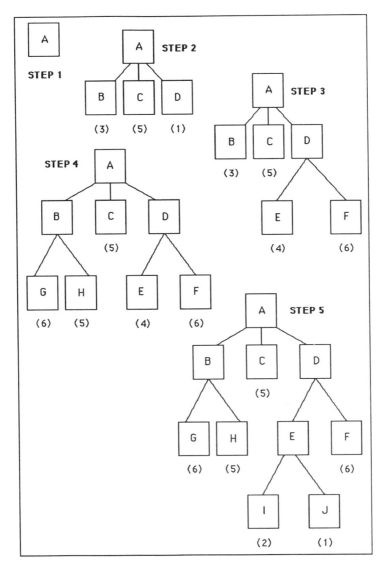

Figure 5.5 Best first search method

This process is continued by expanding further nodes until the destination is reached. This heuristic search method greatly reduces the search effort. However, this method does not always guarantee finding the minimum path.

In the case of a dynamic route guidance system, there is a trade-off between finding a good route quickly and the best route in a longer time. A heuristic search method is therefore required which can find a good route in an acceptable time period. The algorithms used by a number of route guidance systems are described as follows.

THE A* ALGORITHM

The A* algorithm (Hart *et al.* 1968) is similar to the one used by the Philips CARIN system (see Thoone, 1987) in calculating optimum routes. It uses a method based on the best first search approach described in the previous section.

The A* algorithm defines operations on the data associated with each node. These data define the node itself; provide information on how promising the node is, based on the heuristic function employed; describe the link back to the node from which it came; and lists the nodes which can be generated from the node in question.

In the A* algorithm, each generated node is put into one of two sets; open or closed. The set of open nodes contains all nodes that have so far been generated and have had the heuristic function applied to them, but have not yet been examined. The open set is prioritised according to the heuristic function value of each node in the set. The set of closed nodes includes all nodes that have been examined.

The heuristic function of the A* algorithm may be defined as the sum of the cost of getting from the initial state (origin node) to the current state, and the estimated additional cost of getting from the current state to the goal (destination node). This is usually denoted by the equation:

$$f' = g + h'$$

where: f' is the value of the heuristic function

g is the cost of getting from the origin to the current node

h' is the estimated additional cost of getting from the current node to the destination

The value of the heuristic function, f', therefore provides an estimate of the cost of getting from the origin to the destination along a route including the current node. If more than one route is found including a particular node, then the algorithm records the best one.

The actual operation of the algorithm is very simple. It proceeds in steps, expanding one node at each step, until it generates a node that corresponds to a goal state. At each step, it picks the most promising of the nodes that have so far been generated but not expanded. It generates the successors of the chosen node, applies the heuristic function to them, and adds them to the set of open nodes, after checking to see if any of them have been generated before. This guarantees that each node appears only once, although many nodes may point to it as a successor. The next step then begins.

The A* algorithm starts with the open set of nodes containing just the origin node. The value of g is set to zero, since the origin and the current node are coincident. The value of h' is set appropriately. At this starting point, the value of the heuristic function, f', is the same as the value of h', in accordance with the equation shown above. The closed set of nodes contains no nodes at the start of the process. The algorithm then follows the procedures described below until the destination is found.

First, the node in the open set with the lowest heuristic value (f') is selected. This is called the "top node" and is removed from the open set. If the top node is the destination, then the procedure has been completed, and the node and its route from the origin are reported. If the top node is not the destination, then its successors are generated. The value of g for each successor is calculated as the sum of the value of g for the top node and the cost of getting from the top node to the successor.

Each successor is checked to see if it is the same as any node already in the open set. If it is, then the best route from the two available is selected. If this best route is the route calculated previously, then the present successor under examination can be ignored. If the best route is the route using the present successor, then the old route can be ignored. Where the successor has not already been investigated, it is added to the open set.

This process continues such that the origin points to its successors, and each successor in turn points to its successors. This continues until each route terminates with a node that is still in the open set, or a node with no successors. Once the successors of a node have been established, the node can be added to the closed set. The network of possible routes is extended by a series of depth first searches (described earlier). Progressing along a route changes the g value and hence the value of the heuristic function. Each route terminates when either a node has no successors, or it is a node to which an equivalent or better path has already been found.

The A* algorithm can be exploited in a number of ways, based on the method described above. It allows a network to be expanded based on how close the current node is to the destination (h') and how far it is from the origin (g). Using this g value as part of the heuristic function implies that the next node to be expanded will not always be the closest to the destination. The algorithm can be simplified to find a route from origin to destination in a shorter calculation time. This can be achieved by setting the value of **g** to zero. The next node to be expanded is then always the one which is estimated to be closest to the destination.

Another simplification of the algorithm enables a route to be found with the fewest junctions. This can be achieved by setting the cost of travelling from each node to its successor as one. These modifications to the A* algorithm give it the power and flexibility to calculate a good route very quickly, allowing the driver to start the journey, and then to calculate the best route after the journey has commenced.

The second part of the heuristic function is the estimated cost from the current node to the destination (h'). The value of h' is an estimate of the actual cost from the current node to the destination, h. Therefore, if h' is a perfect estimator of h, then the A* algorithm will converge immediately to the destination. Taking the other extreme, where h' is always zero, the search is controlled by the value of g. This produces a search procedure similar to the breadth first search described earlier.

In reality however, the value of h is neither perfect nor zero. By applying the assumption that the value of h' never overestimates the actual value of h, the A* algorithm is guaranteed to find an optimum route to the destination. Applications of the A* algorithm described by Nilsson (1980) take this concept one stage further by defining lower and upper bounds for the estimate of h. This assists in reducing the number of nodes requiring expansion, thus reducing the computation time to find the optimum route.

THE NICHOLSON BI-DIRECTIONAL METHOD

A different approach to the A* algorithm was conceived by Nicholson (1966). This aims to find the shortest route between two points in a network by investigating a selection of routes from both the origin and the destination. The method of route selection is decided by extending the routes which have currently covered the least distance. When a route from the origin meets a route from the destination, a complete route has been found. This needs to be checked to ensure that it is the minimum route. The Nicholson method is used as a basis for some of the systems developed in the Japanese RACS programme.

As with all route choice algorithms, the Nicholson method begins by defining the nodes and links in the network. Each node is allocated a node number, and each link is described by its distance and direction. From a route guidance perspective, this allows for the inclusion of one-way streets. The distance parameter described in this method can be substituted for cost or time according to the optimisation characteristic required.

The objective of the Nicholson method is to find the minimum distance between an origin and a destination, with the least amount of calculation. The method simultaneously examines all the routes out of the origin and into the destination as far as the next nodes. The node which is least distant from the origin or destination is then extended further. This process is repeated until a route from the origin has a node on it which has already occurred on a route into the destination, or vice versa. The method then continues by checking that this complete route is the shortest possible route

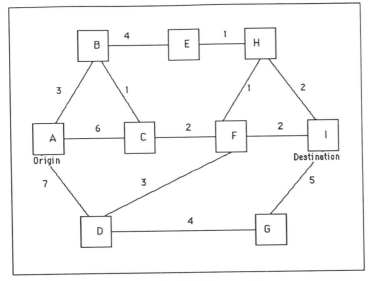

Figure 5.6 The Nicholson problem

from the origin to the destination. The Nicholson method is described by means of an example as follows.

The problem is defined as determining the shortest route between node A and node I, as shown in Figure 5.6. In this example, each link is assumed to be bidirectional. The first step is to expand nodes A and I as far as the adjacent nodes. This is shown in Figure 5.7, with the distance from the origin or the destination shown at the end of the route.

Step 2 of the process examines the result of Step 1. It is noted that the shortest routes so far are H to I and F to I, which both cover 2 units. These two routes are therefore expanded further, as shown in Figure 5.7. The route from H to F (and vice versa) is ignored, since it is longer than the existing method of getting to F. The distances shown in the figure are again from the origin or the destination.

As a result of Step 2, two complete routes from origin to destination have been determined. The route A-C-F-I has a distance of 10 units, and the route A-D-G-I has a distance of 12 units. It cannot, however, be automatically assumed that the route A-C-F-I is the shortest route through the network. There are still some routes in the network which have not yet been examined. If, for example, a connection is found between nodes B and E with a distance less than 4 units, then this will result in a route from A to I of less than 10 units.

The third step of the process continues by examining the shortest routes from the origin or the destination. After Step 2, it can be seen that the shortest routes that have not been investigated are A-B (3 units) and E-H-I (3 units). The first of these is expanded further in Step 3, as shown in Figure 5.7.

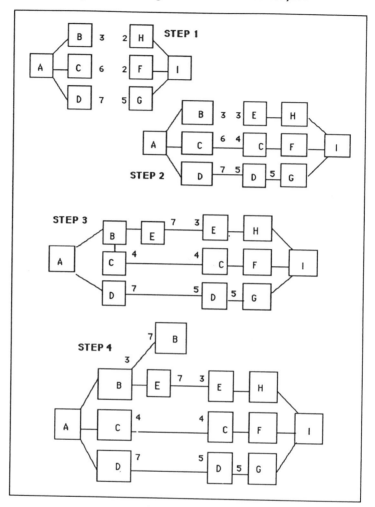

Figure 5.7   Steps of the Nicholson Method

This yields two further connections. The first of these joins B to C in a shorter route than that previously established. This therefore replaces the previous route to C. The second connection joins B to E, creating a third complete route through the network. Three routes have now been established from the origin to the destination. These are A-B-E-H-I, A-B-C-F-I (replacing the former A-C-F-I) and A-D-G-I, with overall distances of 10, 8 and 12 units respectively.

The minimum distance of an unextended route from A is now 4 units, and from I it is 3 units. This implies that a route through the network could be found with a distance of 7 units, thus reducing the existing minimum of 8 units.

Step 4 applies the Nicholson rule again and extends the shortest route from the origin or the destination. This is the route E-H-I (3 units). The extension is shown in Figure 5.7, indicating that no additional routes are established.

The minimum unextended route from the origin is now 4 units, and from the destination it is also 4 units. The minimum route potentially available from the unextended routes is therefore 8 units, which is not less than the existing shortest route. The shortest distance is therefore 8 units on the route A-B-C-F-I.

The algorithm required for the Nicholson method requires two steps. The first step selects a route for extension and extends the route. The second step checks to see if the shortest route has been found. The steps are repeated in turn until the condition of the second step is satisfied. Route guidance and traffic control are described by Nicholson as the two main applications of the method.

RACS ALGORITHMS

This section describes the types of algorithm developed as part of the Japanese RACS (Road and Automobile Communication System) programme which was described in Chapter 1. These are based on the following six methods:

● Dijkstra

● Nicholson

● node amount equality

● elliptical boundary

● network degeneration

● re-routing.

Each of these conceptual route finding methods is described below.

Dijkstra and Nicholson

The Dijkstra (1959) method of finding the shortest route through a network is similar to the Nicholson method described in the previous section. However, where Nicholson uses a bi-directional search starting from the origin and the destination, the Dijkstra method is a uni-directional search from the origin only. These are represented in simple form in Figure 5.8 (a) and (b).

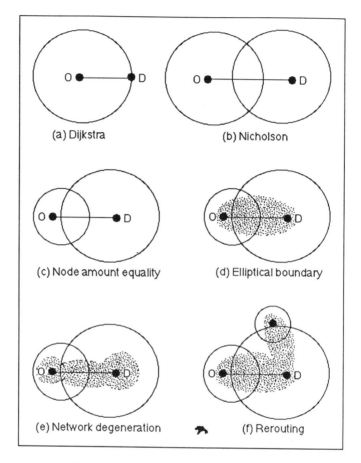

(a) Dijkstra

(b) Nicholson

(c) Node amount equality

(d) Elliptical boundary

(e) Network degeneration

(f) Rerouting

Figure 5.8   RACS route finding methods

The Dijkstra method can be seen as expanding the radius of a circle out from the origin until the destination is included in the circle. Using the Nicholson method, two circles of approximately equal radii are expanded, one centred on the origin and one on the destination. This reduces the area covered in the Dijkstra search by approximately one half.

Node amount equality

The node amount equality method can again be related back to the Nicholson bi-directional method. In the Nicholson method, it is the distance from the origin or the destination that is used to determine which route needs to be further expanded. For the node amount equality method, the number of nodes expanded from the origin or the destination is used.

This results in the network of routes generated from the origin having approximately the same number of nodes as the network of routes generated from the destination. The distance covered from the origin and the destination need not be the same. Hence, in the simple case using circles to represent the extent of the calculated routes, the radii of the circles centred on the origin and the destination need not be the same. This is represented in Figure 5.8(c).

Elliptical boundary

The elliptical boundary method of route finding is based on the assumption that an optimum route is likely to be close to a straight line joining the origin and destination. The method uses the node amount equality method as a basis. An additional weighting is added to each link to reflect how closely it follows the straight line between the origin and the destination. If a link is close to this straight line then its weighting is higher than a link further away.

The resulting distribution of calculated routes is represented in a simplified form in Figure 5.8(d). The ellipse generated by this method has its foci at the origin and the destination. This reduces the number of calculations required considerably in comparison with the node amount equality method. There are two drawbacks with this method, however. The first is that detail of each link is required in relation to the origin and the destination. The second, and more serious concern, is that unlike Dijkstra, Nicholson and node amount equality, the elliptical boundary method is not guaranteed to find the optimum route. It can, however, be used as a good first estimate of the optimum route which can be used to get the driver started on the journey.

Network degeneration

This method aims to improve the efficiency of the elliptical boundary method. It works on the assumption that near the origin and the destination, motorways, main roads and subsidiary roads can be used. However, between the origin and the destination, it is unlikely that subsidiary roads will yield the best route.

The network is therefore graded by road class. In the areas around the origin and the destination, every road needs to be examined to see if it lies on the optimum route. In the middle areas between the origin and the destination subsidiary roads are not considered, thus reducing the number of calculations performed in the elliptical boundary method. This is shown in Figure 5.8(e).

The network degeneration method can be taken one stage further for larger networks, by including only motorways in the area between the origin and the destination when they are a large distance from each other. This creates a considerably more efficient system for calculating routes between two urban areas. As in the elliptical boundary method, this may not always produce the optimum route, but it does provide a good first estimate which can be improved upon whilst the journey is in progress.

Re-routing

The final method of route finding considered by the RACS programme is the re-routing method. This has a specific application when the driver changes from the planned route, or starts from a changed origin. The re-routing method allows the calculations performed from the destination to be re-used in calculating a new optimum route. This reduces the time required to calculate the new route after a deviation from the route by the driver, or changes in traffic conditions near to the driver. The same principle can be applied in reverse when traffic conditions near to the destination change during the journey, but the original route is maintained by the driver.

Table 5.1 Algorithm run times

System	Algorithm	Calculation times (seconds)		
		Small city	Medium city	Large city
Travel-- pilot + RDS-TMC		3	5	6
CARIN	A*	10–30	30–60	60–120
RACS:				
NEC Corp	—	0.002	0.01	0.02
Toyota	—	3	8	25
—	Dijkstra	5	110	245
—	Nicholson	5	20	70
—	Node amount equality	5	18	50
	Elliptical boundary			
	+ Dijkstra	<5	<110	5–245
—	+ Nicholson	<5	<20	5–70
—	+ Node amount	<5	<18	5–50
—	Network degeneration	5	15–20	15–70
	Elliptical boundary			
	+ Network degeneration			
—	+ Node amount	<5	5–10	10–15

When the elliptical boundary method is used to calculate the initial optimum route, some of the routes originally abandoned from the destination near the edge of the ellipse need to be re-opened. The first stage of the rerouting method is to expand from the revised origin until the expansions from this node and the destination are similar. The procedure then continues as before. Figure 5.8(f) represents the calculation areas required in the rerouting method. Again, since this is a modification to the elliptical boundary method, it is not guaranteed to find the optimum route, but will give a good first estimate.

Method combination

The six methods described above were all implemented by RACS programme participants based on the Nicholson bidirectional algorithm. The software was developed on a personal computer with a 2 Mbyte RAM disk, using the C programming language. This allowed any combination of the six methods to be used simultaneously. A map data base was prepared using 2800 nodes and 4200 links based on the areas around Yokohama, Kawasaki and Tokyo.

Comparisons of different method combinations were then carried out. Using a different algorithm for each of the Dijkstra, Nicholson, and node amount equality methods, this showed Dijkstra to be significantly worse in terms of calculation time (Table 5.1). Nicholson and node amount equality were comparable for distances up to about 10 km. Beyond this the node amount equality method was better. The three methods received similar ratings when each was combined with the elliptical boundary method.

Using the network degeneration method for different levels of road classification yielded some significant results. As the number of classification levels increased, the calculation time decreased substantially. However, the larger the number of classification levels, the lower the probability of finding the optimum route.

A system has been developed and tested in the RACS programme using a combination of the node amount equality, elliptical boundary and network degeneration methods. This performs in-vehicle optimum route choice calculations, and has been shown in the pilot study to perform sufficiently well for a dynamic route guidance system.

SUMMARY

There are many algorithms available for finding shortest paths through networks. Those which have been identified as most effective when applied to a traffic routing situation have been described. The A* algorithm is similar to that used by the Philips CARIN system. The Nicholson approach forms the basis for some of the systems developed as part of the Japanese RACS and AMTICS programmes. Some Japanese manufacturers have identified the Dijkstra method as providing the concept for their algorithms.

The A* algorithm can be used in a number of different formats. Each of these is based on the cost of the route currently explored and the estimated cost for the remainder of the route. The Nicholson method expands from both the origin and the destination. The route selected for further examination is determined by the route which has so far travelled the least distance from the origin or the destination. This bidirectional method is guaranteed to produce the optimum route.

The systems and their associated algorithms are summarised in Table 5.1, which gives estimated calculation times, where available. The calculation times should be viewed with caution, as different network assumptions may have been made by the various manufacturers. There may also be differences in the processor power used by the alternative systems, as well as different data storage media, data structures and processor overheads due to support of other functions such as colour graphics.

Although these limitations make detailed comparisons difficult, one general conclusion can be drawn. This is that methods are available for rapid in-vehicle computation of optimum routes, allowing response times of only a few seconds even in large and complex networks. Where necessary, approximations can be made which greatly reduce computation times, giving good, initial recommendations very quickly. These can be verified over the next few minutes, and further refined in those rare cases where a better option is identified.

6 RADIO MESSAGING AND RADIO DATA SYSTEMS

Radio messaging is a digital broadcasting technique used by paging services. It is a system with considerable potential for supporting mobile driver information systems. Because radio messaging is technically relatively simple, it is used here to introduce the technical aspects of digital communications systems in general.

Radio paging is defined by CCIR Recommendation 584 (CCIR, 1982) as a 'non-speech, one-way, personal selective calling system with alert, without message or with a defined message such as numeric or alphanumeric'. The broader potential of pager networks for use as digital communications elements of driver information systems has only been recognized more recently. By broadcasting information to drivers by the car radio it is possible to provide information on road and traffic conditions, either by the traditional method of announcers who interrupt programmes, or by using the traffic message channel on radio data systems (RDS), which will be described in the second part of this chapter.

RADIO MESSAGING

The principle of radio messaging for driver information has been described in some detail by Walker (1990), and this section makes use of his descriptions. The relationships between system components are illustrated in Figure 6.1. Travel information from diverse data sources is relayed to an information centre by any appropriate means of data communication. The information is sorted and collated for transmission via a network of paging transmitters. In-vehicle mobile receivers monitor the pager network and process any messages addressed as traffic information. A thorough discussion of radio messaging, and of field trials for the development of a European standard for transmitting traffic messages is presented by Davies and Klein (1991).

Each mobile receiver contains at least one address (a binary coded number) which it will recognise when received on the pager channel. This

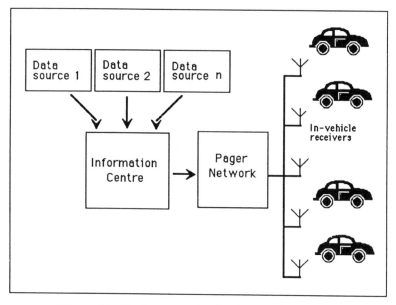

Figure 6.1   Radio messaging for driver information

address identifies the destination and type of information which follows, for those mobile receivers authorised to pick up the data. All mobile receivers accepting traffic information of a certain type will share the same address for this purpose.

In conventional radio paging applications, the pager units each contain a unique address. Each pager recognises its own address when broadcast, and alerts the user by means of an audible or silent (vibrating) alarm. Any digital message which follows the address will then be displayed to the subscriber in numeric or text form. Mobile driver information units may contain such a unique address, allowing personal messaging, as well as one or more general information addresses for traffic and other data.

The traffic information function of mobile driver information units can be classified as a group paging application. Group paging allows a group of users to be contacted with a single transmitted message. Any number of users can be contacted simultaneously, at no additional cost to the service provider. The digital message in traffic bulletins is then decoded by authorised receivers and presented as traffic information to drivers.

A particular traffic bulletin may not be received by all mobiles, or may occasionally be received incorrectly. Service providers must take this into account by repeating the message at reasonable intervals (e.g., 1 to 15 minutes). This also helps to deal with the problem of mobiles newly entering the service area of a particular traffic bulletin. Message updates are also handled by this process of repeat data transmissions.

Messaging networks

Radio messaging has been divided into two broad categories (Tridgell, 1990). These are wide-area systems, and local systems. Local systems cover limited areas such as a hospital, and often utilise a single, low-powered transmitter. Wide-area systems may use several medium or high-powered transmitters, covering a metropolitan area, a region, or a whole country. The wide area systems are of most immediate relevance for mobile driver information applications.

Public, wide-area systems tend to carry a high data load, and use relatively sophisticated modulation and protocols. Paging receivers are often supplied by several manufacturers, and use standard, non-proprietary codes. Traffic bulletins are likely to originate from one or more information centres, and may share the network with any number of other applications, in which personal pager messages typically predominate. Bulletins are grouped and transmitted in batches by the pager network communications controller.

Messaging networks can be installed, operated and expanded at relatively low cost, compared to other mobile information services. They are highly efficient users of the radio spectrum. For example, using the most commonly accepted radio paging code at 512 bits/s, up to eight traffic event messages per second could be sent, each including a problem location, event code, extent, duration and diversion advice.

Where networks are used for radio paging (at an assumed 0.1 calls per pager, and a mixed system with 50% alert-only calls, 37.5% 10-character numeric message calls, and 12.5% 30-character text message calls), a 1200 bits/s continuous system could accommodate some 350 000 subscribers at air-time fills of 80% to 90% (RCSG, 1986). For 90% air-time fills, the mean delay between call input and transmission could be around 3.6 s.

System development

The first radio paging system in the UK was installed in St Thomas's Hospital in London, in 1956. Initially an audio frequency loop was placed around the building, and used to trigger very simple tuned frequency receivers. Later the system was changed to use a 35 kHz carrier modulated by audio tones. Such systems are described as tone coded.

Subsequent systems used much higher frequencies, typically in the range 80–1000 MHz. These permitted the use of integral antennae in the receivers and gave relatively good penetration into buildings. Typically, tone coding was used, with 2 of 30 different tones being sent sequentially to define any one of 870 possible addresses. These capabilities were often sufficient for private system applications, for example within industrial complexes.

The first public, wide-area paging systems were developed in the US and Canada in the early 1960s. At first, callers phoned an operator who keyed in the pager address. Direct dial systems were soon developed , in which automatic equipment checked the validity of the dialled pager address, acknowledged this to the caller, memorised and queued the calls, switched on the transmitters, and sent the waiting calls in a batch. One such example was Bell Canada's System Wide Area Paging (SWAP). In Europe, vehicle-mounted pagers were launched in Benelux (1964) and Switzerland (1965).

With increased demand, two-tone coding gave way to five-tone coding, creating an address capacity of 100 000 users. However, tone-coded systems are prone to errors, because there is little redundancy in the signal. Also, the circuitry in receivers is not well suited to designs using low-cost digital integrated circuits. Moreover, the code is limited and unsuitable for complex messages such as alphanumeric text. Thus, binary coded receivers began to be produced in the late 1960s.

In the UK, British Telecom opened its first paging service in 1973. A national system was initiated in 1976, based in London. At first, the system used two different proprietary codes, transmitting each code alternately in batches. Later, an industry standard code was created by calling together manufacturers from all over the world.

This Post Office Code Standardisation Advisory Group (POCSAG) reached agreement on a code containing full numeric and alphanumeric message capabilities. The POCSAG code was adopted as a world standard by CCIR in 1982, under the title of CCIR Radio paging Code No. 1 (RPC1). The RPC1 code is further described below.

Another form of radio paging, with numeric message capability, was developed in Sweden in 1978. Here the paging code is carried as an RDS (Radio Data System) message, together with other kinds of information. RDS utilises a 57 kHz subcarrier on broadcast FM radio. This has the advantage that separate paging transmitters are not needed. However, penetration into buildings may be less than that desired for a paging service. RDS pagers automatically tune over the broadcast channels until clear paging reception is obtained. This makes them inherently more complex than is necessary for single frequency paging systems. RDS paging is currently used in the US and in Canada, as well as Sweden.

Message reception

At low frequencies, radio waves diffract (bend) around large objects quite well. However, at paging frequencies of 80 to 1000 MHz, there is little diffraction and a deep radio shadow occurs behind any obstruction such as a hill, a building, a vehicle or a human being. Shadowing by buildings causes very patchy reception in built-up areas. However, these frequencies reflect off most hard-surfaced objects, causing many local shadows to be filled in

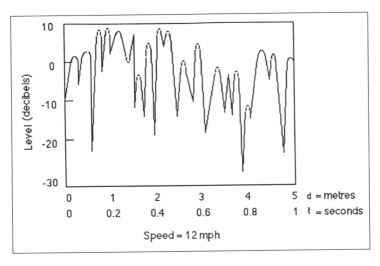

Figure 6.2 Typical multipath fading (at 900 MHz)

by reflections. In urban areas, therefore, reception nearly always consists of several reflected rays, and is termed multipath.

The most significant obstruction is the radio horizon, which results from the Earth's curvature. This is about 30% farther from the transmitting antenna than is the visible horizon, due to refraction in the upper atmosphere. Beyond a transmitter's horizon, signal strength falls off very quickly, allowing the same frequency to be reused without interference.

With a higher transmitter, the distance to the radio horizon is increased. The area (not radius) covered is approximately proportional to the antenna height of the transmitter and receiver. Between the transmitter and horizon, in open, flat country, the received power reduces approximately with the inverse fourth power of distance from the transmitter.

Multipath propagation has the property that, while multipath rays generally add to the strength of the received signal, at some points they subtract from one another so that the signal is severely weakened. These weak points are termed nulls or fades, and occur at about half-wavelength intervals. Fades tend to be up to 20 dB weaker than the average field strength, but fades of up to 40 dB below the local mean level are not uncommon (Figure 6.2). The combination of shadowing and multipath produces a radio field that varies considerably over short distances, giving up to 60 or 70 dB difference in field strength at street level within a 100 m square.

Noise and interference

Reception quality will suffer in any radio receiver tuned to a specified channel or frequency, if unwanted radio power present in the channel approaches or exceeds the power of the wanted signal. The reduction can vary from slight loss to total obliteration of the wanted signal.

A fundamental law of physics is that all receivers generate electrical noise (i.e. unwanted radio power within themselves). The designer's task is to make this noise as small as practicable, but there is a limit below which reduction is not possible. If the power level of the received signal is comparable with the receiver noise, reception will be impaired. This sets a noise limit to the sensitivity of the receiver.

Unwanted power can arise outside the receiver from another transmitter on the same channel (co-channel interference), or from transmitters on other frequencies combining to produce power within that channel (intermodulation interference). Moreover, the channel selectivity of a receiver cannot be perfect, and a nearby transmitter tuned to an adjacent channel can interfere. Finally, man-made noise (e.g. from poorly suppressed engine ignition systems) frequently causes interference.

In laying out any radio transmission system, the task of the system designer is to ensure that the wanted received power exceeds the unwanted noise and interfering signal powers over a high percentage of the intended coverage area (e.g. 99%).

Multi transmitter systems

The basic service that mobile radio offers is communication with a person, or vehicle, at any location. The recipient may be out in the open and on a high hill, or deep inside the lower levels of a building in a valley. As a person travels, their reception situation varies enormously.

As already described, the physical laws governing radio propagation are such that to cover the whole of a wide area with great reliability is not usually possible from a single transmitter site. Thus, several widely spaced transmitter sites are needed, and a wide-area system must be engineered

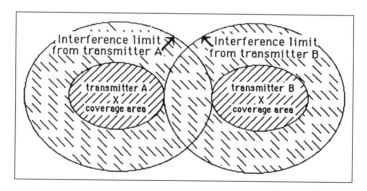

Figure 6.3   Co-channel coverage and interference areas of transmitters

so that these transmissions do not mutually interfere. The techniques of wide-area transmission are concerned with the avoidance of interference.

Figure 6.3 shows a pair of widely separated transmitters, located such that co-channel interference cannot occur. Where radio fields overlap, there are three principal methods of avoiding interference:

(1) *Frequency separation.* Different frequency radio channels are used for adjacent transmitters.

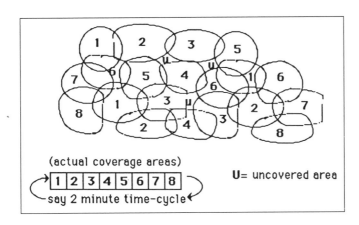

Figure 6.4   Layout of sequential paging transmitters

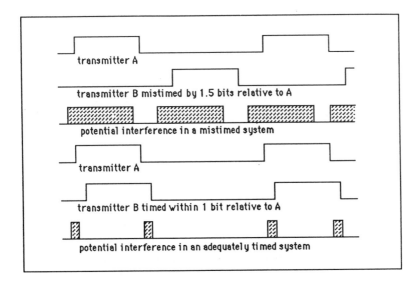

Figure 6.5   Potential interference in a simulcast binary Frequency-Shift Keying (FSK) system

Alternatively, using the same channel frequency for all transmitters:

(2) *Sequential transmission.* While any one transmitter is being used, all surrounding transmitters which might interfere are switched off. Thus, the transmitters are switched on in turn (see Figure 6.4).

(3) *Simulcast (or quasi synchronous) transmission.* The modulation signals of all the transmitters in the area are controlled so that they are substantially in synchronisation. Figure 6.5 illustrates the potential effects of unsynchronised and synchronised digital signals from two transmitters, interference being possible whenever the two signals differ. The actual degree of interference at any location depends on the relative field strengths of the transmitters at that point. Complex analogue signals, such as speech, are much more difficult to synchronise than low-rate digital ones. Typically, speech signals must be synchronised to within 40 ms, whereas 500 bits/s digital signals can tolerate up to 500 ms of mistiming, the latter tolerance being much easier to maintain.

Automatic delay equalisation in a simulcast system

The requirement for simulcast transmission is that the modulating signals be as close to synchronisation as possible at all transmitters. An easy way to ensure this would be to use radio links between the control centre and each transmitter, but shortage of radio spectrum may make this impractical.

If lines are used for linking a control centre to the transmitters, any rerouting by the line provision authority is liable to disturb the line transmission times and the simulcast set-up. Such disturbance may not be detected until complaints by pager users start to accumulate. Further, packet-switched data networks (PSDN) are now available in many areas, but they cannot be used for conveying paging information to base stations unless the inherently variable transmission times of PSDN are taken into account. Thus, automatic equalisation of the delays in a simulcast system is often desirable. Two methods of automatic equalisation are described here; the first is for continuous lines only and the second can include a PSDN.

Method 1   Continuous lines

The paging signals traverse a tapped delay equipment before being distributed to each line and transmitter. Given that the line delay is **TL**, and the tapping delay is *TD*, the object is to adjust each tapping so that

$$TL + TD = TK \quad \text{(constant for the system)}$$

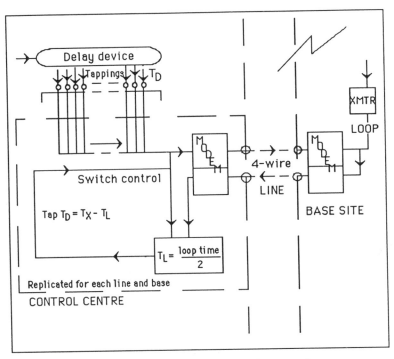

Figure 6.6 Automatic delay equalisation for a simulcast system

Referring to Figure 6.6, it can be seen that the signal for any line is looped back at the base site, and that the returned signal is (frequently or continuously) compared in time with the outgoing signal. The loop delay is divided by two to yield the one-way delay, and the measurement result is then used to select the appropriate delay tapping. Although the time comparator has been shown as individual to each line, because lines do not frequently alter in transmission delay (as compared to the time required for the automatic measurement), a single comparator can be time-shared between all the lines.

Method 2 On-air synchronisation

This method assumes that, apart from a master transmitter, transmission occurs in bursts (i.e. up to several tens of seconds per burst). Each slave transmitter is switched off between bursts, and is equipped with a receiver that can receive the master transmission only while its own transmitter is off. The start of transmission of each slave is derived from the master by this reception. Transmission of the remainder of the burst is then under control of an accurate local clock at each transmitter.

It is possible for one or more of the slaves to act as a submaster and pass on another synchronisation signal to further slave transmitters out of the reception range of the original master. In Finland, the whole national radio paging system is co-ordinated into a single simulcast area by this method. Because timing is passed along the transmitter chain, there is a significant time difference between transmitters in Helsinki and those in the far north, but because these are too far apart to interfere, there is no adverse effect.

We note that interference should not occur over more than about 0.25 of a direct binary FSK signal element. If the frequency accuracy of each local clock (including the master) is $\pm A$ to allow for both master and slave inaccuracies, then:

$$A < 0.25/2TB$$

where TB is the total bits per transmission burst.

For a rate of 1200 bits/s and a transmission burst of 30 s, for example, the accuracy required of the transmitter site clocks is about ± 3.5 parts/million.

Transmission zones

To achieve frequency re-use, large paging systems are sometimes organized into zones. Each zone is covered by multiple transmitters with simulcast transmission. Users pay only for the zones in which they wish to be paged, and a user's calls are transmitted in each of the chosen zones.

Figure 6.7   Radio paging zones and time slots

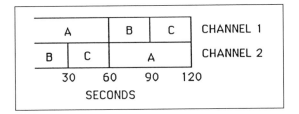

Figure 6.8   Two-channel system

Figure 6.7 shows the arrangement used by the British Telecom (BT) national system. There are 40 zones, each covered by simulcast transmission from several sites. Pager transmissions of the various zones differ due to user zone selection. To avoid interzone interference, the zone boundaries and transmitter sites are planned so that each zone can be allocated to one of three sets, A, B, or C, with the proviso that all zones in any one set are non-adjacent and non-interfering. Consequently, by dividing channel time into three slots and allotting each set of zones its own slot, and by switching on only the transmitters in the slot appropriate to that zone, all interference is avoided and the frequency is reused.

An important property of this air time slotting is that it is geographically infinitely extensible, and thus is particularly suited to the sharing of a single channel across the boundaries of two or more abutting paging systems.

Finally, if two radio channels are available, a single transmitter can be used with remotely switched dual capability at each site. Thus, the cost of a two-channel system can be almost the same as that of a single-channel system (Figure 6.8).

Wide area systems

Within the system coverage area, a location with a weak signal from one transmitter is likely to have a strong signal from some other transmitter. Single-channel systems have a marked advantage over a frequency-separated system in that the receivers are always tuned to all of the transmitters. Thus, the transmissions mutually reinforce each other. The transmissions of a frequency separated system offer no such mutual reinforcement.

In a frequency-separated system, the receiver must be equipped with automatic channel tuning in order to receive throughout the whole area. Tuneable receivers are more expensive, deplete their batteries more quickly, and are more bulky than fixed-channel receivers. Moreover the automatic tuning circuitry must decide constantly whether it should search for some other channel, and communication cannot occur while tuning is in progress.

The throughput of a simulcast system is equal to the maximum possible for a given data (baud) rate, and is unaffected by the number of transmitters in the system. The system synchronisation needs careful set-up and maintenance, but this can be done automatically. A sequential transmission system is easier to set up and maintain. However, for a given baud rate, the paging throughput is inversely proportional to the number of time slots in the transmission cycle. Thus, the achievable efficiency or radio spectrum usage is much less than that of a simulcast system using the same data rate.

A system using sequential transmission can be changed to a simulcast system as the paging traffic increases. Because the same pagers are used, there is no need to disturb customers at the time of this change.

FM radio messaging

Usually, digital messaging channels are dedicated to a single, paging service. The transmitters can thus be planned and sited to give any desired degree of coverage (e.g. penetration into the heart of selected large buildings). The full cost of transmitters is then borne by the system provider. In contrast, one can carry messaging signals on another service, particularly FM broadcast radio. The transmitters are already in place and no separate ones are needed. However,

(1) The receivers need to incorporate demultiplexing circuitry to separate the messaging signals from those of the main service.

(2) The coverage given by the broadcast transmitter network could well provide a lower probability of message reception and therefore a possibility of inadequate service.

(3) Careful business agreements with the broadcast (or other service) organization are necessary to maintain the messaging service to the satisfaction of the users.

Generally, sharing with a broadcast system is cost effective in areas of low population density if indifferent messaging coverage is acceptable.

Multiplexing stereophonic broadcasting and paging

Figure 6.9 shows the standard FM broadcast stereophonic baseband signals with an added data channel. The data signals have a rate of 1187.5 baud, this being time-locked to the 1:48 to 57 kHz subcarrier frequency. This subcarrier frequency is phase-locked to the third harmonic of the 19 kHz stereophonic pilot tone. The data channel is recommended by the CCIR. Data are conveyed in 26-bit blocks (16 information bits and 10 redundancy bits), which

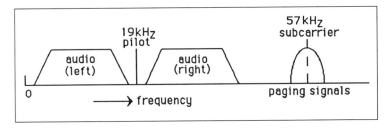

Figure 6.9 Multiplexed paging and broadcast baseband signals

permit correction of any one error burst up to 5 bits long. One class of data catered for is radio messaging.

A pager for this multiplexed band sweeps continuously across the whole FM sound broadcast until it finds and locks onto a good channel carrying data and paging signals. The whole FM signal is demodulated, and then the audio channels and pilot tone are filtered to leave only the data signals. These are then demodulated to recover the 1187.5 baud data.

The information coding can carry several types of data such as road and traffic condition information as well as paging. The unwanted information must be ignored.

Message coding

Messaging calls and similar digital transmissions occur in coded form. The code is selected for ease of transmission, accuracy, and ease of reception. Originally, audio "tone coding" was used, but this is now obsolete.

Binary coding is becoming prevalent, using frequency shift keying (FSK). With this approach a positive shift of radio frequency represents a binary 0 and a negative shift a binary 1. The bits are transmitted in continuous streams or blocks. These transmissions consist of the information bits and some check or parity bits. This permits the received information to be checked for accuracy and for a few errors to be corrected.

Unfortunately, slow movement through a multi-path fade can produce a long burst of errors. In this situation the amount of error checking that needs to be added to permit correction of such a burst is considerable, and greatly reduces the information throughput of a channel. For all those situations within a coverage area that receive a good signal, this massive redundancy is not required and is a waste of transmission time. Therefore, codes must be carefully chosen to achieve a balance between being unable to correct a high proportion of received blocks because too little error checking is included, and wasting transmission time because the checking is excessive for most situations.

Coding requirements

The minimum requirements of a modern radio messaging code are as follows:

- adequate address capacity (especially in an expanding market);
- transmission speed adequate for the intended traffic;
- high protection against false calling;
- suitability for sequential and simulcast transmission;
- low-cost decoding implementation within a pager;
- adequate sensitivity;
- some protection against multi-path propagation effects, but all codes will still suffer some effects;
- acceptable to more than one manufacturer (for wide-area codes);
- numeric and text message capability (essential);
- adequate protection against false message generation;
- compatibility with other codes (important when a code is introduced to an already working system).

Binary coding

The addresses and messages are nearly always transmitted in N-bit codewords. Some codes use one codeword per address and others use two. One code alters both bit rate and codeword size when proceeding from addresses to messages.

Each codeword includes its own fixed number of redundancy bits, usually according to a cyclic code, which permits simple decoding with a good degree of error detection. For a given quantity of redundancy (as fixed by the code), some is used in the pager for error correction. The amount remaining for error detection is thereby reduced, and so the protection against false decoding is weakened. The amount of redundancy needed to achieve error correction rises rapidly with the amount of correction chosen. Thus, code selection is a compromise between traffic throughput and redundancy.

The cyclic block code treats the information as a binary number followed by the same number of zeros as wanted redundancy bits. This is divided (modulo-2) by a generator number (i.e. binary division without carrying because of simplicity of implementation in digital circuits), and the remainder from this division becomes the number of redundancy bits so that the codeword is an exact multiple of the generator. To detect received errors, the received information and redundancy bits are divided by the generator number. If the resultant remainder is not zero, errors have occurred.

The radio paging code no. 1

The RPC1 is summarised as an example of a binary code. The code has a binary format with normal transmission rates of 512 or 1200 bits/s using direct FSK modulation. This bit format permits a synchronisation tolerance between neighbouring transmitters of 488 or 208 ms for the respective rates. The signalling rates are different from those of proprietary codes, and this permits pager battery saving in systems that transmit a mixture of codes.

Each transmission starts with a preamble of at least 576 bits of alternate 0 and 1. The preamble is followed by a string of 32-bit codewords, each with 21 information bits and 11 redundancy bits. For RPC1 codewords, at least the following error protection algorithms are possible:

(1) Using "hard" decision decoding, detect any 5 random errors or any 1 error burst up to length 11, correct any 1 error and detect up to 4 errors or 1 error burst up to length 7, correct any 2 errors and detect up to 3 errors, or correct any 1 error burst up to length 4.

(2) Using "soft" decision decoding, correct any 1 burst up to length 11 including poor quality bits, correct any 5 poor quality bits, or select either according to the quality pattern.

The codewords are transmitted in contiguous batches, each starting with a unique synchronisation codeword and followed by 16 information words. Each batch is thus only 1.0625 or 0.453 seconds duration according to bit rate. The transmission can then be stopped. Thus, time-slotted multi-transmitter systems can be engineered very conveniently. Priority calling within about 1 s can be included in nontime-slotted systems.

The first bit of each information word is a flag to indicate whether the word is an address or message codeword. (Twenty bits are available for information.) The last two information bits of each address codeword convey the call function. The first 18 bits form part of the pager identity, and each pager thus has four addresses. Typically, these are indicated by four different beep patterns.

For the transmission of pager addresses, a batch is partitioned into 8 frames of 2 codewords each. The frame number is part of the address. This multiplies the address capacity of the code by 8 to give over 2 million identities and 8 million addresses, permits further battery saving of about 82%, and reduces the false calling rate by a factor of 8:1.

Any message immediately follows the pager address. Two types of message format are standardized, numeric and text. The numeric format packs 5 digits into each message codeword. Each digit can be any of the decimals or space, or hyphen, or an urgency symbol, or a bracket, or a spare character. The text format utilises the full CCITT Alphabet No. 5, which is also known as ASCII or ISO 7-bit. This may be very important in the future for linking pagers to "lap-top" computers.

A unique "idle" address word is specified to fill in any part of a batch containing no information. (The synchronisation and idle codewords differ by 16 bits, but have a very simple relationship, which can reduce pager implementation costs.)

Radio pager technology

Figure 6.10 shows a typical radio pager design. It is powered by small batteries which last for several months of normal use. The designer aims to achieve great sensitivity together with very low energy consumption.

Typically, a loop or small ferrite rod antenna is integral. This rather inefficient antenna (i.e. 20 dB less sensitive than a dipole) feeds the front end, which selects the radio channel, amplifies, and recovers the modulation waveform from the radio frequencies. The front end is a major power consumer and is automatically turned off whenever possible to prolong battery life.

Multi-frequency systems use discrete component circuitry, which makes them both bulkier and more expensive than single chip solutions. Radio pager front ends tend to use about 5 mW of power, which makes them rather susceptible to electrical noise. The digital parts of the pager tend to be electrically noisy, and the designer has to be careful that these parts do not detract from the front end performance.

An alert-only radio pager usually performs all data interpretation and management functions on a single, specially designed, integrated circuit

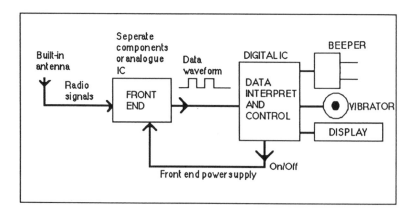

Figure 6.10   Typical radio pager design layout

(IC) digital chip. Message display pagers tend to use a small general-purpose microprocessor and a storage chip. The radio pager address is often held in a small read-only memory (ROM), which can be changed as required (e.g. to suit change of ownership).

The main functions of the interpret and control part of the radio pager are:

- to derive bit timing from the recovered data waveform;
- to synchronise with the data frames;
- to interpret (and perhaps correct) the data;
- to search the corrected data for any of the radio pager's addresses;
- to generate beep patterns or drive the vibrator;
- to memorise any received calls when switched to the memory code;
- to display any stored messages;
- to manage all battery energy saving.

Radio message origination

Public radio message systems usually interconnect with the Public Switched Telephone Network (PSTN). Many PSTNs use Dual Tone Multi-frequency (DTMF) tone dialling, which permits continued use of the keypad once the call has been connected to the radio paging controller. Thus, a common dialled number can be used for all messages, and the wanted pager address and any numeric message can be dialled after the controller answers the call. Various proposals have been made for entering text messages from numeric keypads, but none of these seems to have achieved widespread use.

It is possible to connect radio paging controllers to telex and data networks, or PSTN data via data modems, and so allow those with access to keyboard machines to input text messages directly. Otherwise, text messages are input via a manual bureau, such as a telephone answering message service (TAMS). A wide-area radio paging controller may contain:

- a file of permitted pager addresses with the services and areas allowed for each pager;
- a dialled number file if dialled numbers bear no simple relationship to the pager addresses;
- interconnection to the PSTN (and perhaps other networks);
- encoding subroutines to transmit several different paging codes sequentially;

- batching arrangements for the calls;

- transmitter power-up and power-down control;

- paging code output arrangements;

- a paging traffic statistics package;

- automatic changeover of faulty lines, transmitters, etc;

- a fault diagnostics and reporting programme;

- a system configuration package;

- a billing output, sometimes in bill format, but more usually as a log of calls for off-line processing.

Current status of radio messaging

The US is the largest radio paging market. For wide-area paging, two types of pagers are popular, the tone-only and the tone-plus-voice. However, the advent of message display pagers, combined with a shortage of paging channels, is making considerable inroads into the tone-plus-voice market.

US paging systems are fairly typical of those found with a modern PSTN. Call input is via the telephone DTMF keypad to a paging controller. Recorded voice confirms receipt of a tone-only call. A synthesised voice reads back to the caller any numeric message input via the keypad, or the PSTN automatically provides the radio paging controller with the caller's number so that it can be used as the numeric message. Text message calls are input either via a manual bureau or by some form of data transmission. Transmission to the radio transmitters generally is via UHF radio links where frequencies are available because this is cheaper than leased lines, and simulcast time equalisation requirements are eased.

Tone-coded pagers are being replaced by binary digital types using either the Motorola GSC or the CCIR RPC1 code. There were estimated to be well over two million wide-area pagers in use in the US by the late 1980s.

In Japan, radio messaging was introduced in 1978 with an NEC code. This permitted only 65 000 addresses and was transmitted at 200 bits/s (compare to RPC1 with 8 million addresses and transmission rates of 512 or 1200 bits/s). It proved insufficient for many cities and extra channels were allocated. However, Japan recognized the spectrum-saving virtues of a higher-speed code with much larger address capacity, and RPC1 is now being introduced. There are at least 1.5 million pagers in Japan. Radio messaging is extensively used in Hong Kong, Singapore and South Korea, with the RPC1 code dominating the market growth.

Both Australia and New Zealand have RPC1 paging systems, the latter having a nation-wide system. In Australia, apart from main towns, the population density is too low to support paging. Some form of direct satellite paging system might be a solution to the low population density problem in Australia and other countries.

In Europe, the UK leads the radio messaging field. The BT system is reputed to be the largest in the world (Tridgell, 1987), with over 400 000 pagers and growing at about 3000 per month. The BT system uses the RPC1 paging code (at both 512 and 1200 b/s), and includes five paging controllers linked by data transmission lines. As well as BT, there are several other competing systems. Facilities are similar to those in the US, except that DTMF keypads are not in common use and there is no automatic advice of the calling number. Where DTMF is not available for input of numeric messages, the caller can use a wired or acoustically coupled DTMF keypad.

In France and Germany, the main paging system is Eurosignal, which uses tone-coded amplitude modulation, and was designed circa 1971. The maximum calling rate that any transmitter can handle is 0.6 tone-only calls, and there is no display message facility. Moreover, pagers must be manually or automatically tuned. The total number of such pagers is thought to be about 170 000. However, in Germany city-wide systems using RPC1 are now being allowed to compete with Eurosignal. Neighbours Austria and the Netherlands both use RPC1 in their national systems.

In the Nordic countries, paging is a fast growing service. All except Sweden use only RPC1. Finland has developed an interesting system of simulcast over the whole country. In Sweden, the main system is shared with the FM broadcast radio, being invented and operated by the telecommunication administration. However, a competing, dedicated paging channel RPC1 system is now being constructed.

For international paging in Europe, plans are maturing to provide a common radio message channel at 466 MHz to cover significant parts of France, Italy, the UK, and Germany. The code will be RPC1, transmitted at 1200 baud. In the UK, the channel will be operated by a consortium of paging interests. A much larger pan-European system, known as ERMES, is under development. Studies are being conducted to define a 16-channel system using frequency-agile pagers at about 169 MHz to permit roaming throughout all participating European countries. Market studies forecast about 13 million users by the end of the century.

The European radio message system (ERMES)

In 1987, CEPT (the European Committee of PTTs) conceived the idea of instituting a paging and messaging system to cover Western Europe. A study group, called "RES4", was set up to produce the specifications. Subsequently RES4 was incorporated into ETSI (European Telecommunications Standards Institute) and renamed "PS"; it is charged with completing the specification work. Currently, the system is referred to as the European Radio Message System or ERMES, but the latter name may not be acceptable in all countries.

The services and facilities of ERMES have been fully defined, but much work remains to be done on the definitions and specifications of the radio and networking aspects. The action plan is to complete these definitions within ETSI-PS, and for an ETSI project team to then write the specifications. The service is expected to be operational in 1993.

Dimensions of ERMES

The dimensions of ERMES were derived by considering the forecast number of paging users, the expected calling rates and message volumes, and the number of common radio channels that can be made available throughout Europe. An additional complication is that each country will provide its own system (or systems) on its own radio channels. Studies by CEPT members and marketing consultants have been combined to provide the following dimensions:

(1) The number of users is expected to be about 13 million by the end of this century, and may double before the market is saturated. (Note: The total population of Western Europe is about 350 million.)

(2) The area with the greatest traffic density will be where France, Germany, and the Benelux countries share borders. In that area, at peak hour, the net user traffic could be 20 kbit/s, 25 kbit/s, and 15 kbit/s, respectively.

(3) A block of eight 25 kHz channels at 169 MHz was designated for 1992, to be followed later by another similar block (i.e. 16 channels in total). Of course, not all of these channels will be available at the border of any one country.

(4) From (2) and (3), the peak net user bit rate per channel should be 3.75 kbit/s, although a rate as low as 3.0 kbit/s might be acceptable. Allowing for error-checking bits and system overheads, the eventual channel transmission rate should be in the range of 5.0 to 6.5 kbit/s.

(5) Up to 5% of users will ask for roaming. Not all of these will have their roaming facility simultaneously activated.

ERMES: User and caller facilities

Any user will be able to specify the area in which they want their calls to be transmitted for a defined period, and the pager in that area will adjust automatically to this. Additionally, should the user so wish, the caller will be able to nominate the desired area in which the call is transmitted. As a result, the pager must be able to select the appropriate channel (or channels) wherever the user roams.

To avoid possible loss of calls while the user is in a coverage overlap situation, any call transmitted in an area adjacent to the user's home area will also be transmitted in the home area. If there is more than one network covering the chosen roaming area, the roaming network used will be chosen by the home area's network operator.

Ascending categories of calls are tone-only, numeric message (maximum 20 digits), and text message (maximum 400 ASCII characters). The highest category that can be received by a pager is the designer's option, but the pager must also be able to receive every lower category.

The remaining category of call includes a transparent data message (maximum 4000 user bits). Because there is no standard format, the pagers for this category will be special. The category could be important for dealing with paging developments (e.g. highly compressed digital speech).

Generally, networks have the option of which call controls they offer. Authentication and legitimation will be used to ensure that call control does not fall into unauthorised hands.

There are three call priorities:

Priority 1   This is transmitted within the home area within one minute.

Priority 2   With a mean busy hour transmission delay of less than two minutes.

Priority 3   Which has no defined transmission delay.

Also:

- with user permission, a network may provide a message user directory;

- the caller may be informed of the call charge;

- the caller will receive an acknowledgement of each call made;

- the caller may be informed of the times when calls were transmitted;

- users may bar calls for a specified time;

- both caller and user may request that transmission of calls be deferred until a specified time;

- a user may divert calls to another pager;

- calls may be confined to members of a closed user group;

- generally, the caller will pay the call charge, but a user may opt to pay instead;

- a call may be repeated automatically after five minutes;

- the network will number every message call, and pagers may indicate if any message number is missed; networks may store the messages for up to 24 hours for retrieval by the user, or may offer to transmit the latest message number;

- a message encryption service may be offered by using special pagers;

- calls may be to individuals, to previously specified groups, or multi-addressed;

- a call may be transmitted with an "urgent attention" indication;

- a network may hold a list of standard messages to which the caller can refer by number.

Each pager will have capacity for at least eight addresses (or radio identity codes, RICs) and corresponding alerting signals. A "silent node" can be activated by the user, but will be overridden by any urgent message. In the silent node, a tone-only pager will be able to store a call on each of its activated RICs.

Normal message pagers will store at least ten messages of each type that they can receive, and will also indicate if a message is repeated. The storage properties and capacity of transparent data pagers are not defined. All pagers will indicate the presence of stored calls, and also whether the call memory is full. If the message memory is full, the earliest message will be discarded in favour of an incoming one. Additionally, pagers may offer various message manipulating and reading facilities, an "out of range" indication, a lost message indication, date and time of call indication, and remote programming of RICs.

Technical status

The incomplete status of the scheme has already been mentioned. The following details, however, are available.

Channels: an early decision in RES4 was that all pagers would have the same roaming capability. Consequently, all pagers will provide multi-channel reception. Two schemes were considered:

(1) Local channels would be provided as necessary for local calls. Additionally, a single, time-shared channel would be provided throughout Europe for transmission of roaming calls. Pagers would have two channels only (i.e. their "home" local channel and the common channel).

This simple arrangement required only an extra crystal and crystal selector in the pager.

Commercial and political difficulties, however, were said to make operation of the common channel a dubious proposition, and the scheme was abandoned.

(2) The decision was made that all pagers should be capable of selecting any one of 16 contiguous channels. A synchronised time cycle of about 7.5 s, consisting of 16 equal periods, will be established throughout ERMES. Pagers are divided into 16 groups, and a call for any one group will begin only in their time slot on the pertinent channel.

Pagers situated within range of their home area will select the home channel. Pagers not within range of their home area will, in turn, select each channel for the duration of their group period of that channel. The scheme requires inclusion of a frequency synthesiser in each pager to perform the channel selection, which will significantly increase pager costs, at least initially.

Modulation: several types of modulation have been considered. A transmission rate of about 6 kbit/s was determined to be too high for any two-level type of modulation after taking into account the need to avoid adjacent channel interference, and to be able to use quasi-synchronous transmission. Eventually, four-level FSK was chosen.

Quasi synchronous timing: simultaneous transmissions from adjacent co-channel transmitters will be synchronised to within 10 ms. Transmitters will be sited so that, at a pager, the timing difference between received signals of approximately equal power from adjacent transmitters will not exceed 50 ms.

RADIO DATA SYSTEMS

One method of conveying information to drivers is via the car radio. Traditionally, broadcasters have interspersed radio programmes with traffic news. This service provides motorists with information on road, traffic and weather conditions, which they are unlikely to receive by any other means. The effectiveness of this approach, however, is limited by restrictions on the frequency of programme interruptions and the coverage of broadcasts.

Alternative methods of broadcasting traffic information via radio have been developed within Europe. One of the most significant advances is the Radio Data System (RDS), already mentioned in Section 2. Within this section, the developments which led to RDS are briefly outlined, followed by a description of the main features of RDS.

A development of RDS is the Traffic Message Channel (TMC), which will enable traffic messages to be carried digitally and silently without necessarily interrupting the audio programme. A specification of TMC is also provided in the present module.

Development of RDS

The concept of providing additional information on a subcarrier along with the stereo VHF–FM broadcast is not new. In the mid-1960s, many FM broadcast stations in the US were conveying background music for restaurants or shops on subcarriers. The technique, referred to as "storecasting", was not used by European broadcasters because of unacceptable cross-talk.

In Europe, the German ARI (Autofahrer Rundfunk Information) system was devised in the 1970s as a traffic information broadcasting system. It is a relatively simple tone-signalling system, requiring only a basic decoder. By the mid-1970s, a data-modulated subcarrier could provide much greater information capacity and flexibility.

Radio broadcasters using ARI are required to provide additional information by means of three tones broadcast on a 57 kHz subcarrier to the audio signal. The first of these tones (SK) identifies the radio station as one which regularly carries traffic messages as part of its audio programme. The second tone (DK) is added when traffic messages are currently being broadcast. The third tone (BK) identifies to which of six predefined areas the messages relate.

ARI receivers therefore need to be able to identify these tones, which are inaudible to conventional car radios. The system developed by Blaupunkt in the early 1970s has now been implemented in Germany, Austria, Switzerland and Luxembourg. ARI receivers allow drivers to tune to stations broadcasting the ARI SK tone. Drivers can then mute the radio volume or listen to a cassette. When traffic information is being provided on the radio station, the DK tone is broadcast; this can be used by the receiver to restore the radio volume to a pre-set level, or interrupt the cassette. The BK area tone can be used to allow an advanced ARI receiver to tune to a station covering the area of importance to the driver. In Germany ARI has proved popular, with an estimated 80% of car radios equipped to receive the ARI tones. A similar system, known as ARI-2, has been implemented in a number of states in the United States, although this has not yet gained widespread acceptance.

To build on the success of ARI, an enhanced system has been developed in Germany, known as ARIAM (Autofahrer Rundfundk Information Aufgrund Aktueller Messwerte). This aims to reduce the delay between the occurrence of congestion and reporting of that congestion to drivers by automatic incident detection. The ARIAM system has been described by Schneider (1985) and Giesa and Everts (1987).

ARIAM uses sensors on the highway to detect weather and traffic conditions. Information from these sensors is processed at a control centre and automatically compiled into messages for broadcast as part of the ARI service. Initial tests have shown that ARIAM can provide information to the driver 10 to 15 minutes faster than systems relying on non-automated methods.

The recognition of the need to develop an international standard for radio data systems in Europe, led to the formation of a working group under the auspices of the European Broadcasting Union (EBU). The RDS specification prepared by the group was unanimously agreed by all EBU members in 1983.

A set of functional requirements for RDS were established by the EBU working group (Ely and Jeffrey, 1990). These are as follows:

(1) The radio data signals must be compatible; they must not cause interference to the reception of sound programme signals on existing receivers or to the operation of receivers which use the ARI system.

(2) The data signals should be capable of being received reliably within a coverage area at least as great as that of the main programme signal.

(3) The usable data rate provided by the channel should support the basic requirements of station and programme identification, and provide scope for future developments.

(4) The message format should be flexible to allow the message content to be tailored to meet the needs of individual broadcasters at any given time.

(5) The system should be capable of being received reliably on low-cost receivers.

Features of RDS

RDS is a subcarrier system which provides a silent data channel on existing VHF–FM radio programmes. It represents an ultimate replacement for the ARI system. Both are broadcast on a 57 kHz subcarrier but where ARI offers a simple three-tone system, RDS provides a data stream of 1187.5 baud.

The primary purpose of RDS is to identify radio programmes and allow self-tuning receivers to automatically select the strongest signals carrying those programmes. There are several other applications of RDS which are available. The applications can be split into the tuning functions (programme identification, programme service name, and type of programme), traffic station announcement identification (traffic programme flag and traffic announcement flag), other information (such as time and date, decoder identification and broadcaster applications), radio paging and traffic message channel. Most main European radio programmes now provide the primary RDS features.

In order to offer a service which can be used to replace ARI, RDS provides Traffic Programme (TP) and Traffic Announcement (TA) features. The Traffic Programme feature is comparable to the ARI SK tone, in that it is used to identify radio programmes which regularly provide traffic information as part of the audio programme. The TA feature is then used to indicate when traffic information is currently being broadcast, as with the ARI DK tone.

These traffic-related features of RDS have already been implemented for many local radio programmes. New ARI receivers use ARI and the RDS TA and TP flags to provide drivers with the ARI service. The main traffic information service using RDS, however, will be the Traffic Message Channel (TMC). This will provide a continuous stream of digitally coded traffic information to in-vehicle TMC receivers which will trigger a speech synthesiser or text display. The digital messages will be language-independent, based on a European standard for the TMC feature. The Traffic Message Channel is described in more detail in the next section.

Traffic message channel

The TMC facility of RDS has been the subject of considerable work over the last few years. The work has concentrated on the development of a European standard, addressing the coding of traffic messages, and issues relating to event location and message management.

In the standard development, three main objectives were identified. These were that the coding structure should be efficient, comprehensive and flexible. Efficient coding enables the limited capacity of RDS-TMC to be used to the best advantage. Comprehensive coding ensures that the full range of European traffic situations can be addressed. Flexible coding allows different situations existing between various countries to be addressed and allows the system to adapt to future changes.

One of the earliest methods for coding traffic messages for RDS-TMC was developed by Philips within the Dutch RVI (Radio Verkeers Informatie). This study started in 1985 under the direction of a Traffic Information Steering Group comprising the police, radio broadcasting authorities and the Rijkswaterstaat. The objectives of the study were to develop and implement a prototype RDS-TMC system, including experimental transmission of traffic information over RDS, and field evaluations.

The Philips coding required the use of four RDS groups or sequences for each traffic message. A major conclusion of the group responsible for the study was that a more compact data coding structure was required. This would reduce the probability of error in the data transmission and make more efficient use of the limited RDS capacity.

Work by Blaupunkt on a coding structure for the RDS traffic message channel began at the end of 1985. The structure was based on research by Bosch and used the RDS transparent data channel. The research involved an analysis of the basic structure and content of several hundred messages used in the ARI system. Locations were studied and counted using a detailed road atlas of the area under consideration, in order to determine the size requirements for location coding. This research was subsequently combined with that performed by Braegas and Mardus of Bosch, Hildesheim. The result was the development of an initial coding structure in August, 1986. This comprised at least three RDS sequences per traffic message.

Philips and Blaupunkt began working together in 1986 to produce a more efficient message structure, leading to the joint Blaupunkt–Philips proposal for a traffic message format. The structure of the Blaupunkt–Philips traffic message coding proposal was based on a standard two-sequence message. Messages containing more detailed information could also be defined by using additional RDS sequences. This was the first time that reasonably comprehensive and systematic coding proposals had been drawn up for RDS-TMC.

The Blaupunkt–Philips coding proposal was presented to the European Conference of Ministers of Transport (ECMT) Working Group on Road/Vehicle Communications at the beginning of 1987, and formed the basis of the Madrid proposal. The proposal for defining the allocation of RDS-TMC data fields made to the ECMT in Madrid is set out in ECMT CM(87)8. This proposal was based on the use of two sequences of 104 bits. Within these two sequences, 32 bits per sequence would be used for actual traffic data, giving a total message length of 64 bits.

A communication protocol was devised by the CARMINAT project. CARMINAT is a EUREKA project involving Renault, Philips, Sagem and the French broadcasting organization, TDF. The P3 protocol, concerned with RDS-TMC coding, was essentially an experimental implementation of the ECMT Madrid proposal, combined with a number of message management functions.

From early 1987, Bosch and the Bundesanstalt für Strassenwesen (BAST) worked together to reduce the message length so that most messages would fit into a one sequence structure. The BAST one-sequence proposal shortened the 64 bit content of the two-sequence TMC messages to a minimum of 33 bits.

Further refinements to the coding proposals have been made within the DRIVE programme. A consortium, involving Castle Rock Consultants, the British Broadcasting Corporation (BBC), the Centre Commun des Etudes Télédiffusion et Télécommunications (CCETT), Philips, Bosch and the Transport and Road Research Laboratory (TRRL, now renamed the Transport Research Laboratory or TRL), worked with the EBU and ECMT to define new standards for RDS-TMC. The proposed ALERT (Advice and

Problem Location for European Road Traffic) pre-standard, builds on the previous message coding proposals. It adopts standard European traffic message event descriptions which reflect broadcaster and traffic authority priorities determined through an extensive consultation process within the DRIVE project. Details of the ALERT C Protocol developed by the consortium are presented in the next section.

The ALERT C Protocol

Application layer

The ALERT C Protocol (which will be referred to as "the Protocol") defines the Traffic Message Channel (TMC) as a travel service digitally and silently broadcast using RDS, which can provide a mobile user with the following:

- event-oriented driver information on the nature, severity and probable evolution of both urban and inter-urban traffic problems;

- reduced frustration and uncertainty by providing timely and helpful information;

- assistance with journey planning, including re-routing and rescheduling of trips to avoid current or projected traffic delays;

- details of local traffic incidents which may be avoidable through the use of minor diversions;

- in the future, status-oriented information on traffic conditions which can help to support intelligent in-vehicle route guidance equipment;

- additional data on roadside amenities and tourism information which can, in the future, complement and update in-vehicle mobile data bases.

Information which is broadcast digitally and silently can only be interpreted by suitable RDS-TMC receiver/decoders. These TMC decoders provide the user interfaces with the TMC service. Their function may vary substantially depending on technical developments and market requirements, which cannot be wholly predicted in advance. Instead, a virtual terminal model is defined which covers a range of actual decoder functions, including:

- simple decoders with a limited message repertoire and restricted location data base;

- more sophisticated decoders offering full TMC message features and/or a wide range of strategic and tactical location data bases;

- decoders which monitor only a single, selected TMC programme service, and those which employ more sophisticated multi-channel search strategies;

- decoders which are active before the start of a journey, and those which must acquire their TMC data after the journey begins;

- decoders which provide output via speech synthesis and/or visual displays, and those which interface to more sophisticated in-vehicle route guidance equipment;

- in the future, decoders which will offer additional functions and services yet to be defined.

The protocol defines only event-oriented driver information messages. Provision is also made for the subsequent definition of other types of message, such as status-oriented route guidance information, or such other applications as may be desired in future.

Driver information messages are those designed primarily to service an in-vehicle decoder offering information directly to drivers via speech synthesis and/or displays.

Event-oriented messages describe deviations from the normal traffic equilibrium, and include problems such as congestion, roadworks, adverse weather conditions, accidents, ferry delays or cancellations, etc.

Route guidance information messages would be designed primarily to service an intelligent on-board unit, providing traffic-responsive route selection recommendations in real time.

Status-oriented messages describe current travel situations such as highway section fluidity, travel speeds, link travel times, available car parking spaces, roadside amenities, etc. They are intended to support in-vehicle mobile data bases with real-time travel service information.

As with the case of spoken traffic messages, TMC messages cover different geographic areas. A hierarchical data base of event locations is assumed, including the following major levels of information:

A *International level*
This level includes events on major national and international highways carrying significant volumes of traffic from several countries. Information at this level will be broadcast both in the affected country and in neighbouring countries. Vehicles from all European countries should be able to interpret message locations at the international level.

B *National level*
This level includes events on major national and regional highways carrying significant volumes of long-distance national traffic. Information at this level will be broadcast throughout the affected country. Vehicles from

all parts of the country should be able to interpret locations at the national level.

C *Regional level*
This level includes events purely on primary and secondary regional highways. Information at this level will be broadcast throughout the affected region and, where appropriate, into neighbouring regions. Vehicles equipped with the appropriate regional data base will be able to interpret locations at the regional level.

D *Local level*
This final level includes events on urban highways primarily of local significance only. Information at this level will be broadcast on local radio programmes covering the relevant area. Vehicles equipped with the appropriate local data base will be able to interpret locations at the local level.

The Protocol does not address the internal management of traffic messages by broadcasters in respect of geographic relevance. The Protocol assumes that broadcasters will arrange to transmit messages at the appropriate geographic level(s) using existing procedures such as those defined by the EBU Guide-lines on Broadcasts for Motorists.

Message priorities used by broadcasters adopting RDS-TMC should follow the current approach set out in the EBU Guide-lines on Broadcasts for Motorists. In the context of RDS-TMC, the following range of broadcast message priorities can be defined:

- the highest priority for immediate broadcast is the interruption of existing RDS-TMC message cycles and being repeated very frequently;

- tactical information, for non-delayed broadcast through early insertion into RDS-TMC message cycles, with frequent repeats;

- strategic information, broadcast at fixed intervals dependent on RDS-TMC channel capacity;

- background information, broadcast less frequently, when channel capacity permits.

The Protocol does not address the internal management of traffic messages by broadcasters in respect of broadcast message priority. The Protocol assumes that broadcasters will arrange to transmit messages at the appropriate level of priority using existing procedures such as those defined by the EBU Guide-lines on Broadcasts for Motorists.

The Protocol does address the separate question of message urgency within the decoder. This aspect of the Protocol can be used by receiver manufacturers to determine how a decoder will respond when it receives a TMC message.

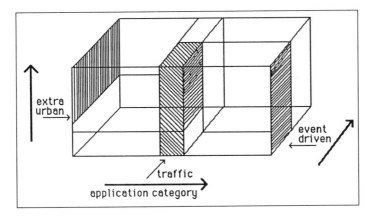

Figure 6.11   RDS-TMC scope

For the purposes of event-oriented driver-information messages, the Protocol utilises a standard European list of weather and traffic problem descriptions at all levels. A short-list of frequently used messages is also standardised throughout Europe.

Provision is made for future extensions of the Protocol on two levels: by using the "grid" concept, and by means of "hooks" within the grid.

The grid comprises a conceptual multidimensional matrix of application elements, each application in principle utilising its own TMC coding structure (Figure 6.11). Each element of the conceptual grid is numbered, beginning with grid element #0 for the event-oriented driver information defined here. The number of the grid element applying to data broadcast by any specific transmitter is indicated within a "service message", defined in the session layer of this specification.

The grid can be used to further define travel services within RDS-TMC and/or to indicate future coding variants or "version numbers" in time or space.

Hooks defined within the current grid element also allow for the provision of future extensions to the Protocol. These hooks comprise binary code combinations left unused in the present Protocol, which are intended to give flexibility in allowing the protocol to respond to changes in the future road transport and broadcasting environments.

A co-operative system of continued liaison is recommended between EBU and ECMT, to permit an orderly evolution of the standard in response to needs agreed at a European level.

Presentation layer

The presentation layer of the Protocol specifies messages which can be presented to the user in accordance with the general requirements set out in the application layer. It defines the message catalogue, its structure and content, and its presentation to the driver.

Traffic Message Channel (TMC) information is conveyed using a "virtual language" in which the codes broadcast over-air comprise addresses of information stored in data bases in the decoders. These data bases contain lists of weather and traffic situations, advice, durations and other information; plus lists of locations, including intersections, road numbers and place names.

Two separate processes are involved in the presentation layer:

- Before transmission, information concerning an event is mapped into the TMC virtual language by selection from nested menus of event descriptions, or by a fully automated traffic monitoring and reporting system.

- In the receiver, the TMC codes are translated into messages using look-up tables.

In this virtual language concept, the message lists used at the source and those used in an individual decoder are not necessarily identical. For example, the messages may be input in one language and reproduced in another. However, within each language or country, standard message lists shall be utilised, whose derivation is the joint responsibility of the broadcasters and traffic authorities of that country.

Much of the information conveyed by the codes is implicit and is derived from secondary look-up tables stored in the decoders. These tables are not addressed by explicit fields in the broadcast information, but are derived from the context of the message itself combined with information from the session layer and other RDS codes already defined by EBU.

The Protocol defines two categories of information within standard messages: basic and optional items. Basic information items are present in all standard messages. Optional information can be added to messages where necessary.

The Protocol also defines short messages, which are available as an option for use by broadcasters faced with very heavy traffic information demands and/or very limited available RDS-TMC capacity. These short messages contain a frequently utilised subset of the basic information items.

Distinction is also made between explicit and implicit information. Explicit information is broadcast directly using defined codes. Implicit information is derived from the secondary look-up tables stored within the decoder, which only occasionally will be explicitly overruled using optional, additional broadcast codes.

Standard TMC user messages provide the following five basic items of explicit, broadcast information (see Figure 6.12):

(1) event description, giving details of the weather situation or traffic problem (e.g. congestion caused by accident) and where appropriate its severity (e.g. resulting queue length);

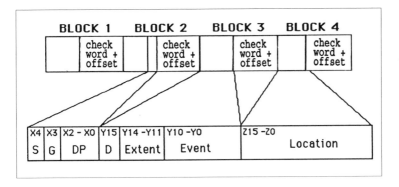

Figure 6.12 Single group standard message

Notes   S   = short message identifier (= 0)
          G   = single group message identifier (= 1)
          DP = duration and persistence
          D   = diversion

(2) location, indicating the area, highway segment or point location where the source of the problem is situated;

(3) extent, identifying the adjacent areas, segments or specific point locations also affected by the incident, and where appropriate the direction of traffic affected;

(4) duration, giving an indication of how long the problem is expected to last;

(5) diversion advice, showing whether or not drivers are recommended to find and follow an alternative route.

Event description (11 bits)

The standard message set comprises an initial repertoire of phrases defined in English. This is structured as specific event descriptions, divided into 24 message types. The message type information is used in the session layer to assist with message management.

Location (16 bits)

The locations utilised by standard messages in the Protocol are defined for each participating country or locality in accordance with guide-lines. The hierarchical location coding principles of these guide-lines constitute an integral part of the protocol. Coding of locations at the international level is the responsibility of ECMT, and at the other levels in the first instance is the responsibility of the national traffic authorities.

Up to 64 sets of locations can be defined for each EBU country code, which forms part of the Programme Identification (PI) code (already defined by EBU Doc. Tech 3244-E). Indication of the location set is provided in the session layer.

Within each location set, problem site location numbers indicate several geographical types:

- country

- region

- area

- highway segment

- point location.

When the source of a problem (e.g. an accident, a bottle-neck) occurs at a defined TMC location, its primary location can be broadcast using the relevant location number.

When the source of a directional problem (e.g. a queue) occurs between two TMC point locations, its primary location can be broadcast using the location number of the nearest downstream point, measured in the direction of traffic affected.

When such an event is determined to be non-directional its primary location can be broadcast using either of the two nearest defined TMC locations which straddle the event.

When a decoder receives a TMC message referring to a location not included in the data base, no message output to the driver will be produced.

Extent (4 bits)

Extent information within standard messages identifies the direction and a location offset of up to seven "steps" through adjacent, defined TMC locations, also affected by the event. The last step in this chain identifies a secondary TMC location which, together with the primary location, straddles the event.

The direction bit (0 = positive, 1 = negative) shall also indicate the direction of queue growth for all incident types defined as directional; i.e., it is opposite to the direction of traffic flow affected. The convention specifying positive and negative directions along each road shall be fixed at the time of coding the definitive locational data base for the country or locality.

When an event affects only one TMC location, the location offset is zero. The direction of traffic affected shall be indicated as above, should the incident be directional.

When, occasionally, events affect more than seven adjacent point locations, they should normally be described at the segment level as being located within one or more segments. If, exceptionally, this is not adequate, further locations affected can be defined using optional, additional information.

Duration (3 bits)

Event duration information in single group messages provides for eight levels of expected continuation of the problem. The interpretation of the duration code depends on the nature of the weather or traffic situation.

For multi-group messages, where the duration field is not present, control codes are defined which have the same effect as the duration field.

Diversion advice (1 bit)

The diversion (included in single group messages only) indicates whether drivers are recommended to find and follow an alternative route around the traffic problem describe elsewhere in the message.

SUMMARY

Digital communications systems, particularly those on radio messaging techniques, have considerable potential for disseminating information to drivers. This discussion started by reviewing the concept of radio messaging to support mobile driver information systems by describing how the messaging networks operate and how the systems have been developed. Details of the technology and potential reception problems have been discussed to indicate the likely performance of the systems.

This chapter has also described the Radio Data System and the Traffic Message Channel. Advances in radio broadcasting have enabled drivers to receive relevant information on current traffic conditions. RDS-TMC has the capability of increasing still further the quantity of broadcast traffic information.

The Protocol has built on previous traffic message coding proposals to define a proposed pre-standard for RDS-TMC throughout Europe. The Protocol includes the message catalogue, its structure and content, and its presentation to the driver. Implementation of RDS-TMC will increase the availability of traffic information at relatively low cost.

The current status of radio messaging demonstrates the demand for these paging services. The ERMES scheme is a very ambitious one. Whether these ambitions prove to be commercially viable is not yet known. However, there is no doubt that the volume of radio paging and messaging data will increase greatly over present levels in the future. Because of the limited radio spectrum available, schemes with high transmission rates will certainly be needed to meet these future requirements. With further developments such as ERMES this technological approach is likely to be one of the main communications systems in Road Transport Informatics.

7 COST–BENEFIT ANALYSIS: A CASE STUDY

This chapter describes a case study which illustrates some of the technologies previously mentioned. The study concerns approximate cost–benefit analyses carried out on three alternative route guidance scenarios for the Amsterdam metropolitan region, for the year 2000, and was conducted on behalf of the Transportation and Traffic Research Division of the Rijkswaterstaat (the Dutch Ministry of Transport). Following this introduction, the three technological scenarios are described. We shall then discuss the semi-dynamic and dynamic benefits associated with each scenario. System costs are then detailed, and the final part of the chapter summarises the analysis results.

THREE TECHNOLOGICAL SCENARIOS

Here we describe three possible scenarios for the implementation of route guidance technology in the Netherlands. These form the basis of comparative cost–benefit analyses presented later in the chapter. The three scenarios are:

(1) route guidance based on self-contained, semi-dynamic on-board navigation systems;

(2) route guidance based on low-cost in-vehicle units with short-range infra-red external communication links to a roadside;

(3) route guidance based on self-contained, semi-dynamic on-board navigation systems, with a long-range communication link to the Radio Data System-Traffic Message Channel (RDS-TMC).

Each scenario has been set in the context of initial implementation within the Amsterdam metropolitan region. The scenarios will now be described in turn and in more detail.

Scenario 1. Self-contained system scenario

In the first scenario, a proportion of vehicles based in the Amsterdam region in the year 2000 would be equipped with on-board navigation systems similar to the CARIN system. As indicated earlier, these can be divided into simple directional aids; location display systems; and self-contained guidance systems. Of these, self-contained guidance systems offer the greatest potential benefit, since they offer actual routing advice to the motorist. These types of systems were therefore selected as the basis for evaluation in the analysis.

Self-contained guidance systems can assist drivers in both route-planning and route-following tasks under normal traffic conditions. On their own, they can take account only of typical conditions and are unable to cope with real-time variations in the actual situation. In order to provide maximum benefit, the map data bases on which these systems work need to reflect network characteristics such as peak and off-peak travel costs and times, as well as distance. The system evaluated is assumed to contain such "semi-dynamic" information.

A characteristic of self-contained systems with on-board data bases is that their information on available routes and journey times tends to get out-of-date. Most systems allow for periodic updates by storing the data bases on CD-ROM, or other exchangeable media. Even so, it is likely that a typical autonomous system will be providing guidance advice based on two- or three-year old data. This factor is considered in the evaluations which follow.

An effective, self-contained navigation system would eliminate a proportion of the excess travel currently incurred through navigational waste. The main benefits of the system would therefore be to individual drivers who bought and utilised the system. However, there may also be some benefit to all traffic: the reduction in congestion caused by the elimination of some of the excess travel incurred by equipped vehicles. Benefits would be gained both within Amsterdam and wherever else the equipped vehicles travelled, as the equipment would work throughout the country and beyond.

The self-contained guidance systems included in this scenario could use dead-reckoning combined with map-matching or a tri-lateration technique as the basis for fixing the vehicle location. In each case, motorists would need to initialise the system by keying in codes for their required destination using a keypad. The system would then compute the best route through the network for the particular time of day using its minimum path algorithm.

Presentation of the routing advice to the motorist could be achieved through a variety of interfaces, singly or in combination. These include alphanumeric displays, graphic displays, speech synthesis units or other audio signals. Visual displays may be located at dash panel level or may take the form of head-up displays, similar to those used in aircraft.

Scenario 2. Infra-red system scenario

The second scenario involves provision of route guidance in Amsterdam using a system similar to those demonstrated in London and Berlin, in the Autoguide and ALI-SCOUT trials (dynamic route guidance systems discussed in Chapters 5 and 6). This would involve a proportion of the vehicle population being fitted with lower-cost, in-vehicle units capable of short range two-way infra-red communication with roadside beacons located at strategic points on the urban road network, and linked to a central computer facility.

The system included in this scenario would utilise a roadside infrastructure consisting of post-mounted infra-red transmitter/receivers. An in-vehicle dead-reckoning and map-matching capability would provide for route-following between beacons. The beacons would be suitable for mounting on existing traffic signals, and the associated electronics would be housed in signal control roadside cabinets which would minimise the cost of the infrastructure.

With this type of system, the two-way communication link enables each vehicle to supply the infrastructure with journey time information on the section of network it has just travelled, as well as receiving information on impedances for alternative routes ahead. This floating car information is used by the central computer to update a changing model of network conditions. This model is used as the basis for supplying information to the in-vehicle units, which then follow pre-calculated routes.

Systems which utilise two-way communication links in this manner, are able to operate in a fully dynamic mode, and take into account changing traffic conditions. The advantage of this type of system over an on-board navigation system which is not responsive to changes in network conditions is illustrated in Figure 7.1, which is taken from the von Tomkewitsch (1987) evaluation of the ALI-SCOUT trials in Berlin. A static on-board navigation system treats determination of the optimum route between two points as a two-dimensional problem, since the impedance along each link between the two points is assumed to be fixed throughout the day. However, a dynamic route guidance system treats the journey between the two points as a three-dimensional problem, with link impedances varying with time as network conditions change.

Measured journey-time information offers another advantage over the historical data base approach of Scenario 1, in that its journey time data base should be up-to-date. However, the balance of advantages between measured and historical data is not all one way; for example, route choice at the start of the morning peak based on early morning measurements of journey times could direct drivers into well-known bottlenecks, which would easily have been avoided using knowledge of typical peak traffic conditions. For this reason, Scenario 2 assumes that a combination of

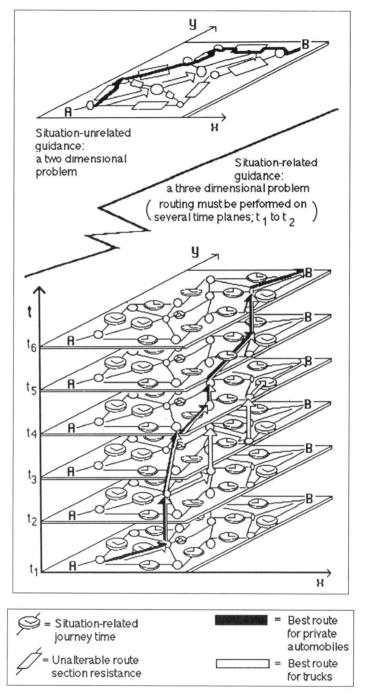

Figure 7.1 The dynamic route guidance concept, after von Tomkewitsch (1987)

measured and historical "semi-dynamic" data would actually be used for route selection.

The degree of responsiveness to changing traffic conditions of the Scenario 2 system depends on the beacon deployment density over the network. In particular, with larger beacon spacing there is a longer "information lag" between a change occurring on a network section, and journey time information for that section being passed to the central system via the roadside infrastructure. For this analysis, it was assumed that roadside beacons are located such that the average journey time between beacons is approximately five minutes.

When a driver begins a journey, the low-cost in-vehicle unit functions only as a directional aid, because no mass data storage is provided giving details of the surrounding highway network. Once a beacon is encountered, pre-calculated routes can be followed. Where recommended routes are not followed, or in areas not provided with beacons, the system provides only very limited, directional assistance, rather than detailed guidance. The key to this system's performance therefore lies in an extensive network of beacons, with beacons located close to the beginning of each trip.

Scenario 3. RDS-TMC system scenario

The third scenario examined in this analysis involves an alternative method of linking an in-vehicle unit to an external traffic information source. In this scenario, on-board navigation systems fitted to a proportion of vehicles would be capable of receiving and processing additional traffic and network information transmitted via the Radio Data System-Traffic Message Channel (RDS-TMC). This data would be used to update the network representation held in the in-vehicle unit both in respect of long-term network changes and short-term incidents, giving a fully dynamic system with a significant degree of real-time responsiveness. RDS-TMC is described in more detail in Davies and Klein (1991), and in Chapter 5.

As in Scenario 1, the on-board navigation system would offer actual routing advice based on stored "semi-dynamic" network data and a minimum path algorithm, which would be presented to the motorist through a variety of media. A map display could be provided as a supplementary information source for drivers (as used in the Bosch TravelPilot or Philips' CARIN systems).

The communications link used to convey real-time network information to in-vehicle units in this scenario is the Radio Data System (RDS) Traffic Message Channel (TMC). The main feature of RDS-TMC is that coded traffic information is transmitted inaudibly, superimposed on normal radio signals. Receivers then interpret this inaudible signal and display a short message to the motorist. The message can also be spoken to the motorist using synthesised speech. Most significantly, however, work is currently in progress on developing on-board navigation systems that can use data input from RDS-TMC to update their network information. It is envisaged that such systems will become commercially available within the next few years.

Unlike Scenario 2, the RDS-TMC supported system would offer full dynamic guidance from the time a trip first started, in addition to which full national coverage could be readily provided. However, the equipment cost would be higher, as we shall see. The trade-offs between cost and performance are complex, and will be further developed through the remainder of the chapter.

The degree of responsiveness of a dynamic system in which an on-board navigation system is linked to RDS-TMC would depend on the sources used to provide information to RDS-TMC for broadcasting. In many countries, these sources are currently limited to manual surveillance by police, road authorities and other agencies, who feed information through to a control centre. However, in Japan, for example, the AMTICS dynamic route guidance system acquires its real-time congestion data from existing systems of detector loops on urban motorways and major surface streets. These monitoring systems and the resulting data have been available since the 1970s, without requiring any additional costs for route guidance data collection.

In the Netherlands, an extensive network of traffic monitoring and incident detection sensors is already in operation on motorways. This would be capable of providing data similar to those used by AMTICS on the Japanese urban motorways. The situation on surface streets in Amsterdam is currently less satisfactory; however, many of the potential benefits of route guidance evaluated in this study would be gained from application of the systems to the future urban motorway network. It is also possible that surface street traffic control systems used in Amsterdam will be improved significantly by the year 2000, giving information comparable to that currently available in other countries.

European systems for responsive urban traffic control on surface streets are already capable of supporting a route guidance system similar to AMTICS, using RDS-TMC for data dissemination. Typically, traffic control systems have loops on each link, either near stop lines or at link entries; parameters such as cyclic flow profiles, saturation flows, congestion and queue lengths can be detected and/or modelled to acceptable accuracies for journey time estimation. The density of loop detectors used by traffic responsive systems is typically much greater than that of infra-red beacons, potentially giving quicker route guidance advice when incidents occur. The information from such systems would therefore provide an excellent basis for RDS-TMC broadcasts, maximising the degree of real-time responsiveness and utility for the system envisaged in this scenario.

BENEFITS OF SEMI-DYNAMIC ROUTE GUIDANCE

In order to assess fully externally linked route guidance, it is first necessary to examine the potential use of an autonomous in-vehicle navigation system, offering "semi-dynamic" guidance using an on-board data base of peak and off-peak travel times. This has been carried out using Amsterdam to represent a characteristic Dutch city. The results derived however, are thought to be generally applicable to many urban areas. This evaluation constitutes the first of the three scenarios being considered. For this stage of the analysis a number of parameters were assumed to remain constant. These are described in the following paragraphs.

A study into the extent of travel time and distance wasted due to poor navigational skills of drivers was completed in 1986 by KLD Associates. This study determined that of all travel, 6.4 % of distance and 12 % of time is wasted. In reality, a self-contained guidance system would be unlikely to recover all of the excess time, due to difficulties in establishing actual minimum time routes. In order to take account of this, a factor of 80 % has been applied to the 12 % figure. This gives a value of 9.6 % for the maximum proportion of travel time that could be reasonably reclaimed under typical traffic conditions using the semi-dynamic route guidance system.

Values of time have been obtained from the Rijkswaterstaat (The Dutch Ministry of Transport) based on research by Cheung, Kleijn and Gunn (1989). Values are available for commuting, business and other use. These were weighted according to the proportions of travel for each purpose at different times of day, obtained from appropriate Dutch data. An average vehicle occupancy level of 1.3 persons was also assumed. A weighted average of the peak and off-peak values was calculated as 18.6 Dutch florins per hour (approx. 8 ECUs or US$ 11).

To perform the analysis, other parameters need to be varied in order to assess the effect this has on the magnitude of the benefits. The parameters to be varied are:

- the proportion of the attainable journey time saved

- the proportion of the attainable journey distance saved

- average lorry and car operating costs

- the average annual time spent in travel.

Having previously established that an in-vehicle route guidance system could result in a 6.4 % distance saving and a 9.6 % time saving, it is necessary to estimate criteria drivers would use to optimise their travel. In many cases the minimum distance route between two points does not coincide with the minimum time route. Drivers will therefore, on average, select a route with some trade-off between time and distance, thereby saving a proportion of excess time, and a proportion of excess distance.

Experience from the Japanese CACS and RACS programmes (now incorporated into the VICS programme), and European investigations into route guidance, suggests that people normally adopt something approaching a minimum time criterion for route selection. This is backed up by work carried out on the ERGS concept. The ERGS study presented survey results suggesting that route choice was made on the following basis:

Least time   63%
Safest route 26%
Most scenic   9%
Least cost   2%

These results indicate that 63 % of the population would follow the "least time" route, and 2 % would follow the "least cost" route. The "least cost" route may not necessarily be the shortest distance, although as a reasonable first estimate this could be assumed to be the case. This would imply that 63% of the attainable time saving and 2% of the attainable distance saving is achieved by the equipped population.

More recent studies (Wotton and Ness, 1989) have also asked drivers about the criteria they use for route selection. These tend to confirm that the majority of drivers (about 70%) aim to choose the least time route, with some 10% aiming for the least distance route. Special criteria apply to scenic routes for leisure travel and designated routes for goods vehicles. The results have been summarised as follows:

	Quickest	Shortest	Scenic	None or Other
Journey to work	76.0%	11.4%	0.9%	11.0%
Business	73.6%	9.3%	3.5%	13.8%
Goods vehicle	68.6%	8.5%	0.8%	22.3%
Leisure	47.9%	10.3%	28.8%	11.7%

These results, however, do not take into account the possibility of the minimum time route and the minimum distance route coinciding. Although drivers may select their route on the basis of minimum time only, this may often be the same as the minimum distance route, and vice versa. Consequently, in reality, the proportions of attainable time and distance saved would be somewhat greater than the 63% and 2% assumed in the ERGS study.

For the present analysis therefore, low, medium and high values for the proportions of time and distance saved have been estimated as shown:

	Time	Distance
Low	65%	15%
Medium	75%	25%
High	85%	35%

The low estimate corresponds to savings of 63 % and 2 % being attainable for time and distance respectively through the driver's route selection, with a small additional benefit in each category brought about by minimum time and distance routes coinciding. Since the major proportion of the population prefers least time routes, a significant coincidence of minimum time routes with minimum distance routes will inevitably lead to a large increase in the proportion of distance saved. This explains the higher estimated growth in distance savings, increasing from 2% to 15%, than in time savings, which are assumed to rise from 63% to 65%. The medium and higher estimates correspond to the higher figures indicated by Wotton *et al.* for non-leisure trips, with increasing coincidence of minimum time and distance routes, as well as possible coincidence with other route selection criteria (most scenic route, designated route, etc.).

A second parameter which can be varied is the proportion of vehicles with route guidance equipment. Again low, medium and high values have been adopted. These are 7%, 14% and 25% respectively. The high estimate of 25% has been obtained from work on the London Autoguide electronic route guidance system (Jeffrey, 1981). The medium estimate has also been obtained from earlier work on the same study. The low estimate has been assumed in the analyses to provide a "worst case" scenario.

A value of the average annual kilometres per vehicle is also required. This has been based on typical figures which indicated 16 000 kilometres/year as an average. This value has been checked for the Amsterdam region using data from the Noord Vleugel model of the Rijkswaterstaat, with which it closely agrees . Estimates of annual time spent in travel have been calculated by disaggregating the average distance travelled by the proportions of travel occurring on each class of road. This led to the calculation of an average annual speed of 49 km/h which, when combined with the average annual distance travelled, implies an annual travel time of 327 hours.

The vehicle operating costs were calculated on the basis of published data and previous studies. The marginal vehicle operating cost was assessed as 0.16 DFl per kilometre, based on projections used in the Noord Vleugel study. This figure was doubled for goods vehicle operating costs, in accordance with assumptions of previous studies.

Finally, it is necessary to apply these data to the vehicle population in the Amsterdam region for the year 2000. Valuable data were provided by the City of Amsterdam on current vehicles driving within the relevant region

(493 000 vehicles/day). The growth index provided by the Rijkswaterstaat is projected as 142, giving a future vehicle population estimate of 634 000 for the end of the century. Other data from the Rijkswaterstaat suggest that 10% of these are commercial vehicles, and 90% are cars. Thus, figures of 570 600 cars and 63 400 goods vehicles were finally used in the calculations. The variable parameters used in the benefit assessment are summarised as follows:

			Time	Distance
A	Proportions of time and distance	1	63%	2%
	saved	2	65%	15%
		3	75%	25%
		4	85%	35%
B	Proportion of vehicles equipped	1	7%	
	with route guidance equipment	2	14%	
		3	25%	

For both time savings and distance savings, 12 different combinations of the various parameters can be developed for the benefit assessment. The total annual benefit to the users of route guidance systems is calculated as the sum of the time savings and the distance savings. These in turn are calculated as follows:

Value of annual = proportion × proportion × average × value of
time saving   attainable   excess time   annual   time
                   time saved   travel time

         × proportion × number of × vehicle
           vehicles   lorries   occupancy
           equipped   + cars

Value of annual = proportion × proportion × average annual
distance saving   attainable   excess   kilometres
by vehicle class   distance   distance
                   saved

         × vehicle operating × proportion
           costs by vehicle   vehicles
           class   equipped

Total vehicle   = (number of cars × car operating costs)
operating costs   + (number of lorries × lorry operating costs)

Table 7.1 User benefits of self-contained route guidance

	1	2	3	4	5	6	7	8	9	10	11	12	13	14	15	16	17	18
group 1	6.4	9.6	2	63	7	570 600	63 400	39 942	4 438	0.16	0.32	18.6	1.3	16 000	327	21 222 822	159 967	21 382 789
	6.4	9.6	15	65	7	570 600	63 400	39 942	4 438	0.16	0.32	18.6	1.3	16 000	327	21 896 562	1 199 751	23 096 313
	6.4	9.6	25	75	7	570 600	63 400	39 942	4 438	0.16	0.32	18.6	1.3	16 000	327	25 265 264	1 999 585	27 264 849
	6.4	9.6	35	85	7	570 600	63 400	39 942	4 438	0.16	0.32	18.6	1.3	16 000	327	28 633 966	2 799 419	31 433 385
group 2	6.4	9.6	2	63	14	570 600	63 400	79 884	8 876	0.16	0.32	18.6	1.3	16 000	327	42 445 644	319 934	42 765 577
	6.4	9.6	15	65	14	570 600	63 400	79 884	8 876	0.16	0.32	18.6	1.3	16 000	327	43 793 125	2 399 502	46 192 627
	6.4	9.6	25	75	14	570 600	63 400	79 884	8 876	0.16	0.32	18.6	1.3	16 000	327	50 530 528	3 999 171	54 529 699
	6.4	9.6	35	85	14	570 600	63 400	79 884	8 876	0.16	0.32	18.6	1.3	16 000	327	57 267 932	5 598 839	62 866 771
group 3	6.4	9.6	2	63	25	570 600	63 400	142 650	15 850	0.16	0.32	18.6	1.3	16 000	327	75 795 793	571 310	76 367 103
	6.4	9.6	15	65	25	570 600	63 400	142 650	15 850	0.16	0.32	18.6	1.3	16 000	327	78 202 008	4 284 826	82 486 834
	6.4	9.6	25	75	25	570 600	63 400	142 650	15 850	0.16	0.32	18.6	1.3	16 000	327	90 233 086	7 141 376	97 374 462
	6.4	9.6	35	85	25	570 600	63 400	142 650	15 850	0.16	0.32	18.6	1.3	16 000	327	102 264 164	9 997 926	112 262 091

1　Proportion of excess distance (%)
2　Proportion of excess time (%)
3　Excess distance saved (%)
4　Excess time saved (%)
5　Proportion vehicles equipped (%)
6　Car population
7　Lorry population
8　Number of equipped cars
9　Number of equipped lorries
10　Car operating costs
11　Lorry operating costs (Fl/km)
12　Value of time (Fl/km)
13　Vehicle occupancy rate
14　Average kilometres
15　Annual travel time (hours)
16　Value of time savings (Fl/year)
17　Value of distance savings (Fl/year)
18　Total user benefit (Fl/year)

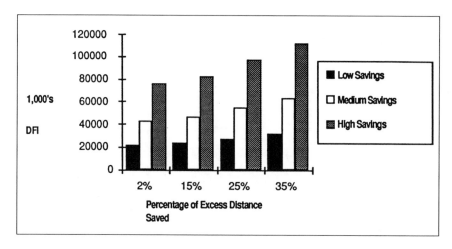

Figure 7.2   Total user benefit in DFl/year

The results have been calculated using a spreadsheet format shown in Table 7.1. Figure 7.2 illustrates column 18 (total user benefits) with a histogram, again showing the savings in terms of thousands of Dutch florins. For ease of analysis, this has been divided into three groups. Each group contains the four pairs of excess time saved and excess distance saved. The remaining information is incorporated as follows:

Average annual user kilometres: 16 000 km

Vehicle operating costs: DFl 0.16/km (car)
DFl 0.32/km (lorry)

Average annual time in travel: 327 hours

Proportion of vehicles equipped: 7% (group 1)
14% (group 2)
25% (group 3)

Analysis of the results shows that the value of annual time saved is considerably more than the value of annual distance saved.

The most optimistic ("best case") scenario gives a total annual benefit to all users of DFl 112 million. The most pessimistic ("worst case") scenario gives a total annual benefit of DFl 21 million. The most likely outcome is estimated as an annual benefit of around DFl 55 million to all users of a route guidance system.

Using these results, it is possible to calculate the average annual benefit to the individual user of an up-to-date semi-dynamic route guidance system. This figure is obtained by dividing the total user benefit by the total number of equipped vehicles. There are four variations on this figure resulting from the combinations of excess distance and excess time. The proportion of vehicles equipped does not affect the average individual benefit, as an increase or decrease in the number of vehicles equipped simply results in a corresponding proportional increase or decrease in the total user benefit. The possible variation in average individual user benefit is shown in Table 7.2.

Table 7.2 Individual user benefits from self-contained semi-dynamic route guidance

Excess distance saved (%)	Excess time saved (%)	Average individual user benefit (DFl/yr)
2	63	482
15	65	520
25	75	614
35	85	708

Table 7.2 indicates that a driver travelling in a vehicle equipped with autonomous in-vehicle route guidance equipment will receive an annual benefit between DFl 500 and DFl 700. These figures allow the driver to weigh the economic incentive of purchasing the necessary in-vehicle equipment against the purchase price of the equipment, with appropriate consideration given to its operating life.

One factor which could tend to reduce these individual benefits is "ageing" of the on-board data base. In estimating the significance of this factor, an analogy has been drawn with the ageing of fixed-time signal plans, which are known to lose 3% of their benefits per year due to changes in traffic patterns. If this figure can also be applied to route guidance systems, average drivers in Scenario 1 could lose 6% of these benefits due to their two-year-old data base. This assumes that drivers will replace their data bases after 3 to 4 years on average.

Some additional benefits would also accrue to vehicles not equipped with the route guidance system. In the London AUTOGUIDE assessment, these additional benefits were estimated as 12% of the user benefit. However, further traffic generation can be expected to reduce the value of this saving, by an amount estimated as 50% in the London study. Thus a net additional benefit of 6% has been assumed, which cancels out the effect of on-board data base ageing.

The potential benefits of Scenario 1 are summarised in Table 7.3. These are the savings due to use of systems without a link to an external infrastructure, providing a semi-dynamic guidance capability. The analysis indicates that very substantial benefits may be gained by helping drivers select and follow efficient routes.

Table 7.3 Potential annual savings from use of self-contained semi-dynamic route guidance equipment

Proportion of vehicles with route guidance equipment (%)	Total annual user benefit million (DFl/yr)
7	21–31
14	43–63
25	76–112

To an extent, these autonomous system semi-dynamic benefits apply to all three implementation scenarios being considered. For Scenario 1, these are the only benefits which can accrue by system implementation. Scenario 2, route guidance using infra-red communication, includes a proportion of these benefits that accrue after the driver has passed a beacon, plus the additional benefits available by updating the route choice information in real time. For Scenario 3, route guidance using RDS-TMC, the full benefits of a self-contained system can be combined with dynamic benefits in order to perform a system appraisal.

There are several factors to consider in determining what proportion of the full, semi-dynamic benefits to apply to Scenario 2. Firstly, the travel speed information received from infra-red beacons should be up-to-date compared to the average two-year-old data assumed for Scenario 1. For this reason, the 6% benefit reduction of Scenario 1 does not apply. Secondly, additional benefits will again accrue to non-equipped vehicles, which can be estimated as 6% of the relevant semi-dynamic benefits.

Two other factors could tend to reduce the semi-dynamic benefits of Scenario 2. The first factor is that whenever an equipped vehicle drives outside the Amsterdam area, it will receive no routing advice at all (unless in the future, beacons are located at five minute intervals all over the country). No reduction in benefit has been assumed from this factor. Instead, the appraisals exclude any differences in benefit incurred outside the Amsterdam region.

The second factor which tends to reduce the semi-dynamic benefits of Scenario 2 is that when drivers start a trip, they receive no routing advice other than static directional aid until they reach their first beacon. Assuming an average journey of 10 km, with 2 km to the first beacon, a driver would be guided for around 80% of a typical urban journey. This figure has been applied to the full semi-dynamic benefits calculated as above.

Similar evaluation of Scenario 3 leads to the following adjustments. Firstly, the on-board journey time data base should be up-to-date, due to night-time downloading of network changes over RDS-TMC. Therefore, as in Scenario 2, no benefit reductions need be caused by ageing. Also, as in the other scenarios, additional benefits of 6% have been assumed to accrue to non-equipped vehicles.

The other two factors considered in Scenario 2 do not apply to Scenario 3. The semi-dynamic on-board navigation system could provide guidance over the whole country, and beyond. Also, when drivers begin any journey, they will receive guidance immediately. For these reasons, the full semi-dynamic benefits have been taken to apply within Scenario 3.

BENEFITS OF DYNAMIC ROUTE GUIDANCE

Fully dynamic route guidance systems offer additional benefits over semi-dynamic approaches, in that they can be responsive to random fluctuations in actual traffic conditions. Random fluctuations may be due to incidents, or simple variations in the numbers of drivers selecting different routes from day to day. The common feature of these random effects is that they cannot be anticipated from historical data alone.

Dynamic route guidance can provide benefits directly to system users and indirectly to non-users. This section considers the overall traffic operations benefits offered by such a system due to real-time response to incidents. In a fully dynamic system, users are directed to avoid the incidents via optimal routes based on real-time data. The current evaluation excludes other random effects, and is therefore only a partial estimate of the total potential benefits of a fully dynamic system.

Scenario 2 considers a route guidance infrastructure based on infra-red communication via beacons, where communication channels are both road-to-vehicle and vehicle-to-road. For Scenario 3, communication is based on the RDS-TMC. Incident detection for this scenario assumes that an automatic incident detection system will have been implemented as part of an advanced traffic control system by the year 2000.

To analyse the overall benefits of full dynamic route guidance for Scenarios 2 and 3, a modelling approach was adopted using a computer simulation. This was designed to assess the potential benefits for all vehicles when a proportion of drivers use traffic information to alter their travel plans.

The model input

The computer model contains two highway facilities. The first (facility #1) represents a motorway or major route, and the second (facility #2) represents all alternative arterial routes combined. The simulation is designed so that incidents can be represented on the major route. A three hour time period is evaluated within the model, with up to three hours afterwards for traffic to disperse in the event of a major incident. The demand during the three-hour period can be made constant or peaked, enabling peak or off-peak situations to be represented.

The model works on a minute-by-minute basis. For each minute, it computes the demand, the effective capacity of each of the facilities, and the speeds on each of the facilities. It then computes the average speed for each vehicle entering (or trying to enter) the corridor during the three-hour period. If demand exceeds capacity, a queue builds up and reduces the effective capacity of the facility.

System benefits are computed relative to a base case. Since some route diversion occurs in the absence of any special information, the model allows for an assumption that a certain percentage of the motorists shift to another facility when they experience heavy traffic. The model allows for three types of response to traffic information: diverting a trip, delaying a trip, or not making a trip at all.

The simulation uses a number of input parameters that can be altered by the operator, as follows:

- maximum demand on facilities #1 and #2;

- capacity of facilities #1 and #2;

- maximum and minimum speeds of facilities #1 and #2;

- distribution ratio—the ratio between the peak demand in the middle of the three hour period and the demand at the start and end of the period;

- length of corridor;

- value of time—Dutch guilders (DFl) per hour;

- operating cost function—the relationship between changes in speed and changes in vehicle operating costs;

- marginal vehicle operating cost per kilometre;

- percentage of vehicles switching from facility #1 to facility #2 in the absence of traffic information;

- percentage of vehicles diverting from facility #1 to facility #2 on the basis of traffic information;

- percentage of vehicles delaying a trip by 1 hour on the basis of traffic information;

- percentage of vehicles delaying a trip by 30 minutes on the basis of traffic information;

- percentage of vehicles cancelling their trip on the basis of traffic information;

- information lag—the time elapsing between the start of an incident and the receipt of traffic information by drivers;

- the direct percentage reduction in road capacity as the result of an incident—the model calculates an effective reduction from this;

- the start and end times of an incident;

- the number of times per year that a particular type of incident occurs.

The main parameters required as input to the model can be divided into the following four categories:

- constant

- corridor-dependent

- period-dependent

- corridor- and period-dependent.

These are described in more detail below.

Constant values

During the analysis for each scenario, a number of the parameters outlined above were held constant. The basis for selection of these constant values is discussed below.

The vehicle operating costs and operating cost function figures used as input data were calculated on the basis of published data and previous studies. An average vehicle operating cost was assessed as DFl 0.17 per kilometre. This is a weighted average of the operating costs used in the semi-dynamic benefit analysis. The operating cost function was conservatively assumed to be 0.13, which lies at the lower end of estimates produced by previous studies.

The net percentage of vehicles that would intuitively switch from the main route to the parallel arterials without route guidance information during build-up of congestion following an incident was considered to be low. The absence of any information on incidents would preclude the majority of drivers from taking appropriate diversionary action. An estimate of 5% of all vehicles was therefore utilised for this input parameter.

The net percentage of vehicles that would delay a trip by 1 hour, delay a trip by 30 minutes, or cancel a trip on the basis of traffic information were all estimated as zero. A driver utilising dynamic route guidance equipment would, upon specifying a destination, be guided along the route calculated as optimal at that moment in time. It therefore seems unlikely that, having been provided with a currently optimal route, a driver would either delay the journey or cancel it completely. These parameters could be used in future analyses to determine the effects of pre-trip planning information.

The information lag is the main difference between Scenarios 2 and 3 in the initial section of the analysis. For a route guidance system based on infra-red communication via beacons, the information lag is essentially dependent upon the beacon deployment density. A high density will lead to a shorter information lag. As indicated earlier, it is envisaged that implementation of the infrastructure in a city such as Amsterdam will result in beacons spaced such that the journey time between beacons is approximately 5 minutes under normal flow conditions. From this, an average information lag can be calculated for Scenario 2.

The average case assumes that an incident occurs mid-way between two beacons. The first part of the information lag consists of the time for a vehicle to travel between the two beacons, through the incident. Since under normal flow conditions this would be 5 minutes, it can be assumed that this first element of the information lag will be greater than 5 minutes; say, 6 minutes for modest delays. The central computers for the Autoguide and ALI-SCOUT systems recalculate link journey times every 2 minutes. So, on average the additional lag in this process will be 1 minute. Finally, once the beacons begin to transmit these revised journey times, a vehicle waiting to receive the information will typically be mid-way between two beacons. This introduces an additional lag of about 3 minutes. In summary, therefore, the total information lag for Scenario 2 will be 10 minutes or more. A value of 10 minutes was chosen for the initial appraisal based on this calculation. Actual information lags under a nominal 5 minute beacon spacing are likely to vary from about 5 to 20 minutes, with the longest times on heavily congested links, and in suburban areas with more widely spaced beacons.

For Scenario 3 a different information lag is expected. This scenario is based on the use of an automatic incident detection system. As with the beacon spacing for Scenario 2, the information lag is mainly dependent upon the detector spacing. Studies by Illinois Department of Transportation in Chicago and by Bosch in Hildesheim indicate that the time between an incident being detected and a message being broadcast on-air will be less than 1 minute. The main lag for this type of system is therefore in the time to detect an incident. For this study, the lag was taken to be 4 minutes, allowing for a queue to build up and for the system to be sure that an incident had occurred rather than just one particular vehicle having stopped. The total lag for Scenario 3 was therefore selected as 5 minutes. Again, a wider range of actual values (perhaps 2 to 10 minutes) can be expected in practice.

It would be possible for a Scenario 2 route guidance system based on infra-red communication via beacons, to also use an automatic incident detection system to reduce the information lag. In this case, in addition to the 4 minutes estimated to detect an incident, there would be the time taken for the central computer to recalculate the link journey times, and to wait for the vehicle to reach a beacon. The resulting delay would be 4 + 1 + 3 = 8 minutes; closer to that of Scenario 3.

Three further differences between Scenarios 2 and 3 cannot be represented by the present dynamic model. Firstly, in Scenario 2, no dynamic guidance is provided until the first beacon is passed, so that drivers initially rely on local knowledge or the limited directional aid system function. Once they have driven some distance in a particular direction, they may be less willing to divert, and have fewer options available. This effect can be reduced by placing the beacons before intersections so that users can select the correct lane and direction at the junction. Secondly, Scenario 2 does not present the driver with any reasons for taking an unusual diversion, whereas Scenario 3 does. This may also affect the driver's willingness to divert. Finally, when serious congestion first occurs, the information lag would suddenly become greater in Scenario 2, as the "floating cars" became held up in queues. These complex interactions are not taken into account within the present analysis.

On the other hand, it can be argued that a central computer (as used in Scenario 2) could possibly make a better estimate of optimal routes than can self-contained systems. This has been represented by the assumption that Scenario 2 will never divert traffic in such a way as to make conditions worse overall, producing negative benefits. All negative results have therefore been replaced by zeros in Scenario 2. Scenario 3 allows both positive and negative outcomes, representing the greater difficulty of centralised planning with a distributed system.

Finally, it is necessary to define the severity of incidents which might occur in the area modelled. Based on work carried out by Lindley (1987) on urban motorway congestion, four incident types were defined (as shown in Table 7.4). Although shoulder incidents have an effect on highway capacity, this is generally only limited and route guidance equipment will therefore provide little benefit in the event of such incidents. Therefore, only incident types #1, #2 and #3 were utilised in the modelling.

Table 7.4 Motorway incident types for a metropolitan area

Incident #	0	1	2	3
Incident type	Shoulder incident	1-lane blockage	2-lane blockage	3-lane blockage
Duration in-lane (minutes)	N/A	25	35	40
Relative frequency (%)	96.0	3.84	0.14	0.02
Relative frequency excluding #0 (%)	—	96.0	3.5	0.5

Corridor-dependent parameters

To undertake a detailed assessment of the potential benefits of dynamic route guidance under Scenarios 2 and 3, the central Amsterdam region was divided into eight corridors. These corridors are shown in Figures 7.3, 7.4, and 7.5, with the main routes shown as solid lines, and the alternatives dotted. Each corridor contains a length of motorway (or main route) and an alternative arterial. Data were provided by the Rijkswaterstaat on link lengths, capacities, peak flows and peak speeds. From these data, a number of corridor-dependent parameters were calculated. These parameters are described in the following paragraphs.

Firstly, the corridor length was calculated by summing the individual link lengths along the main route. Capacities for the main routes (facility #1) and the arterials (facility #2) were determined from the link capacities. The overall facility capacity was calculated as the weighted average of the link capacities, using the link length to provide the weighting. From this capacity data, the number of effective lanes on the main route was determined. This number of lanes, based on the overall facility capacity, was used to evaluate the direct percentage reduction in road capacity as a result of incident types #1, #2 and #3.

Maximum speeds for each corridor required more complex calculations. For each link, the maximum speed was taken as the speed limit. The time taken to travel along each link was then calculated using the link length and speed. These times were summed to find the overall time taken to travel along the corridor. Using this time and the corridor length, the maximum speed for each corridor facility was calculated. The minimum speed is defined as the speed at which traffic can still progress past an incident. The value of minimum speed is largely empirical and was estimated for the eight corridors based on the road types and highway capacities.

Period-dependent parameters

There are two period-dependent parameters required as input to the model. The first of these is a value of time, and the second a distribution ratio. These are both period-dependent, but independent of corridor.

Values of time savings were calculated using data provided by the Rijkswaterstaat, as outlined in the cost–benefit analysis earlier in this chapter. Separate values were used in this part of the study for peak and off-peak appraisals, though the difference is small. The values are DFl 18.3 per hour at peak, and DFl 18.9 per hour, off-peak.

The distribution ratio is defined for the main route as the ratio of the demand at the start and end of the three-hour period to the peak demand during the three-hour period. For peak period analysis, the demand at the

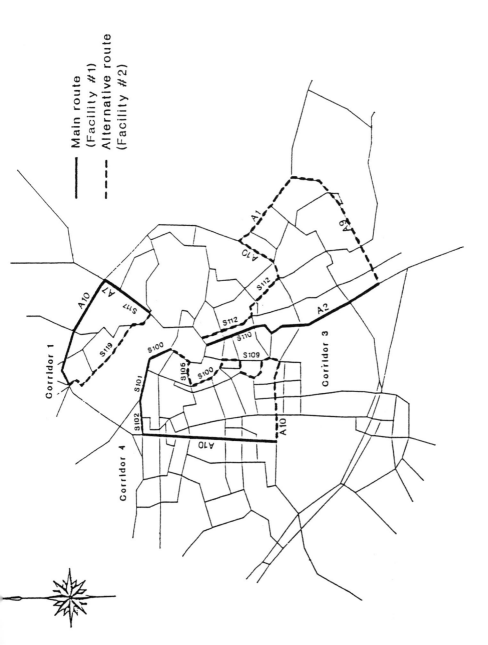

Figure 7.3 Central Amsterdam corridor locations: corridors 1, 3 and 4

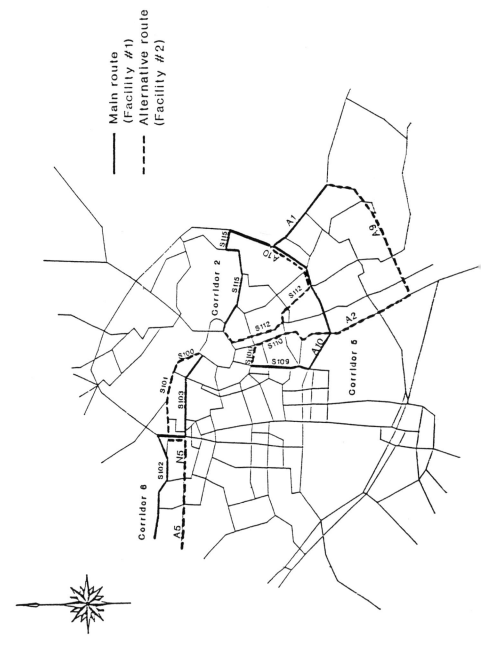

Figure 7.4   Central Amsterdam corridor locations: corridors 2,5 and 6

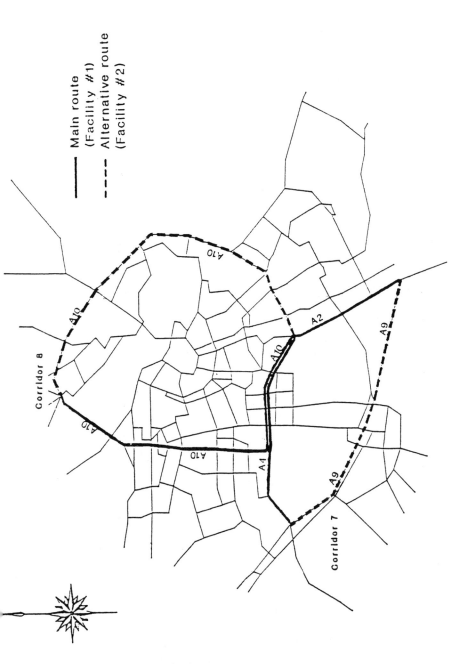

Figure 7.5 Central Amsterdam corridor locations: corridors 7 and 8

start and the end of the period was defined as the off-peak demand, and the peak demand was defined as the maximum demand on the main route. During the off-peak period, the flow was assumed to be constant, leading to a unit distribution ratio. For analysis of the central Amsterdam region, data were provided by the Rijkswaterstaat showing flow distributions at a number of locations around Amsterdam. This was used to determine a distribution ratio which could be applied to the peak period of all corridors under investigation.

Corridor- and period-dependent parameters

In order to calculate benefits from dynamic route guidance the model requires information concerning demands and incident frequencies. These corridor- and period-dependent parameters are described below.

Maximum demands on facilities #1 and #2 were determined for the peak period from data provided by the Noord Vleugel model for each link in the central Amsterdam region. This model, developed by the Rijkswaterstaat, simulates the evening peak for the network shown in Figure 7.3, based on predictions for the year 2000. If the capacity along a particular route in a corridor was the same for each link, the maximum demand was calculated using the weighted average of the link maximum demands, where the weighting was provided by the link length. However, where capacity on a route varied from link to link, the calculations took into account the relationship between maximum demand and capacity. For these cases, a maximum demand to capacity ratio was determined for each link. The value of this ratio for each route was then calculated based on the weighted average method described above. This was then combined with the capacity for the route to obtain the maximum demand during the peak.

Off-peak demands on facilities #1 and #2 were calculated from the peak demands. The distribution ratio, described earlier, provided a ratio of off-peak to peak demand. This was applied to each maximum peak demand to find the maximum off-peak demand.

The final corridor- and period-dependent parameter required as input to the model related to the frequency of incidents. Accident frequencies were provided by the Rijkswaterstaat for the central Amsterdam region. These were combined with data on non-accident, or other incidents. Based on the accident frequencies and the relative frequencies given in Table 7.4, incident frequencies were calculated for each corridor. These were then divided between peak and off-peak periods based on data from the Rijkswaterstaat on accident frequencies for specific lengths of highway. This determined the number of annual occurrences of each incident in a particular corridor during peak and off-peak periods.

Equipped vehicles

The model analyses the effects of vehicles diverting from the main route (facility #1) to the alternative arterials (facility #2). The percentage of vehicles diverting is essentially dependent on the proportion of vehicles equipped with a route guidance system. Three values were adopted for this proportion: 7%, 14% and 25%.

However, in the case of the Scenario 2 route guidance systems being evaluated here, these three figures could not be used directly to represent the percentage of vehicles diverting from one facility to another. This is because, in the event of an incident, a centrally controlled system would automatically optimise the percentage of vehicles diverting rather than divert all equipped vehicles. In the modelling exercise for Scenario 2 optimal percentages of diverting vehicles were therefore determined, up to the three maxima represented by the figures for proportions of equipped vehicles. These optimal percentages were determined through repeated model runs.

Scenario 3 assumed that the in-vehicle software was simpler and did not provide for any kind of system optimisation when recommending alternative routes.

Model summary

The modelling technique for both scenarios utilised the following input:

(1) For all model runs, the following input parameters were set:

- operating cost (DFl 0.17)
- operating cost function (0.13)
- % switching (5%)
- % delaying 1 hour (0%)
- % delaying 1/2 hour (0%)
- % cancelling trip (0%)
- incident start and end times (incident #1, 90-115; incident #2, 90-125; incident #3, 90-130).

(2) Each corridor selected in the central Amsterdam region was considered in turn. For each corridor, the following parameters were set:

- corridor length (km)
- capacity on #1 (veh/h)
- capacity on #2 (veh/h)
- % capacity reduction for incident types #1, #2 and #3
- maximum speed on #1 (km/h)
- minimum speed on #1 (km/h)
- maximum speed on #2 (km/h)
- minimum speed on #2 (km/h).

(3) The analysis was divided into peak and off-peak periods. The following parameters were independent of corridor but were period-dependent:

- value of time (DFl/hr: 18.30 in peak, 18.90 in off-peak)
- distribution ratio (0.35 in peak, 1.00 in off-peak).

(4) The following parameters were set, according to both period and corridor under consideration:

- maximum demand on #1
- maximum demand on #2
- frequency of incident types #1, #2 and #3

(5) For Scenario 2, the information lag was initially set at 10 minutes. For Scenario 3, it was initially set at 5 minutes. Other variations of the scenarios were also examined in the final system appraisal.

(6) Using the appropriate corridor-dependent and period-dependent parameters, the three percentages of vehicles diverting were utilised to establish the benefits accrued. For Scenario 2, an optimal percentage was determined. For Scenario 3, the percentage of vehicles equipped was used as the percentage of vehicles diverting.

The model output

The model output indicates the value of route guidance information to each driver passing through a particular highway section during a three-hour period in which an incident occurs. The total benefit is obtained by first calculating the total number of vehicles passing through the section during the period.

During the three-hour peak period, the total flow is calculated by assuming a parabolic flow, as shown in Figure 7.6. The total flow (TF) is obtained using the equation:

$$TF = F \times (r + 2)$$

where F is the peak hour flow and r is the distribution ratio. The equation is based on an integration of the general parabola equation. The total three-hour flow is the sum of the flows on facilities #1 and #2 in each corridor. During the off-peak, the flow is assumed to be constant, and so the total three-hour flow is three times the sum of the hourly flows on facilities #1 and #2.

The total annual benefit for each corridor for peak or off-peak conditions is then given by the equation:

$$\textit{Total annual} = \text{benefit/vehicle/year} \times \text{total no.}$$
$$\textit{benefit} \qquad\qquad\qquad\qquad\qquad\qquad \text{of vehicles}$$

ANALYSIS OF RESULTS

This part of Chapter 7 concludes by examining the benefits of a dynamic route guidance system to all road users. The results are analysed by taking into account the proportion of vehicles equipped with route guidance, the peak and off-peak periods, and the conditions specific to each corridor.

Peak period analysis

During peak periods the highway network is at its most congested. The use of dynamic route guidance during this period can therefore potentially offer the greatest benefit. The results shown in Tables 7.5 and 7.6 illustrate that route guidance would be of substantially greater benefit during the peak period than during the off-peak.

Closer examination of the peak results in these tables yields a number of findings. Firstly, it can be seen from the results that for most of the corridors selected, route guidance would generally provide a substantial benefit to all drivers. Also, as the proportion of vehicles equipped with route guidance increased (up to the highest level examined in this study), the benefit would increase. The fully dynamic benefits shown in the two tables are generally similar, showing that the different information lags assumed are generally not of major importance. However in one corridor (No. 8) the projected difference is much larger, for reasons discussed below.

The characteristics of each corridor provide further information on when and where route guidance would be most effective. The maximum benefits

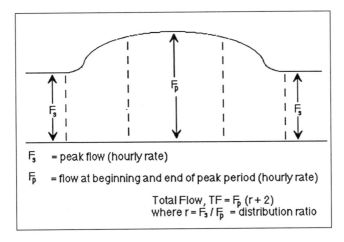

F_s = peak flow (hourly rate)

F_p = flow at beginning and end of peak period (hourly rate)

Total Flow, $TF = F_p (r + 2)$
where $r = F_s / F_p$ = distribution ratio

Figure 7.6   Parabolic distribution of flow

Cost–benefit analysis: a case study

Table 7.5 Annual savings in Scenario 2 (10 min information lag)

Corridor	Percentage of vehicles equipped (%)	Annual savings (DFl 000s)		
		Peak	Off-peak	Total
1	7	0	1.6	1.6
	14	0	1.6	1.6
	25	0	6.4	6.4
2	7	77.4	2.0	79.4
	14	138.6	4.0	142.6
	25	166.8	7.2	174.0
3	7	0	0.4	0.4
	14	0	0.4	0.4
	25	0	0.8	0.8
4	7	900.2	12.4	912.6
	14	1 539.0	15.6	1 554.6
	25	1 959.2	28.4	1 987.6
5	7	607.4	34.4	641.8
	14	3 686.0	67.2	3 753.2
	25	8 950.8	105.2	9 056.0
6	7	46.2	1.2	47.4
	14	80.2	2.4	82.6
	25	131.4	4.0	135.4
7	7	3 639.2	6.8	3 646.0
	14	7 329.2	13.2	7 342.4
	25	9 398.4	20.0	9 418.4
8	7	0	9.6	9.6
	14	0	18.8	18.8
	25	361.6	32.0	393.6
Totals	7	5 270.4	68.4	5 338.8
	14	12 773.0	123.2	12 896.2
	25	20 968.2	204.0	21 172.2

would generally be obtained where the main route has limited spare capacity and the alternative route is relatively uncongested. Three types of incident severity were considered in the corridors modelled. These were a one-lane blockage (type #1), a two-lane blockage (type #2), and a three-lane blockage (type #3). Of these three, 96 % of incidents were one-lane blockages. During the peak period, the capacity reduction caused by a type #1 incident was often sufficient to cause demand on the main route to be over capacity. Where surplus capacity on the alternative arterial was available, route guidance was able to offer substantial benefits.

Table 7.6 Annual savings in Scenario 3 (5 min information lag)

Corridor	Percentage of vehicles equipped (%)	Annual savings (DFl 000s)		
		Peak	Off-peak	Total
1	7	0	2.0	2.0
	14	0	0.0	0.0
	25	0	0.4	0.4
2	7	80.2	2.0	82.2
	14	140.2	4.8	145.0
	25	196.4	7.6	204.0
3	7	0	0.4	0.4
	14	0	0.4	0.4
	25	0	0.8	0.8
4	7	901.8	13.2	915.0
	14	1 656.4	16.8	1 673.2
	25	2 402.8	29.6	2 432.4
5	7	475.0	34.8	512.8
	14	2 453.2	68.4	2 521.6
	25	7 840.4	108.8	7 949.2
6	7	48.0	1.2	49.2
	14	90.2	2.4	92.6
	25	138.0	4.4	142.4
7	7	3 848.2	6.8	3 855.0
	14	7 883.0	13.2	7 896.2
	25	12 439.0	20.8	12 459.8
8	7	2 678.6	9.6	2 688.2
	14	5 129.0	18.8	5 147.8
	25	5 723.4	32.8	5 756.2
Totals	7	8 031.8	70.0	8 104.8
	14	17 352.0	124.8	17 476.8
	25	28 740.0	205.2	28 945.2

Corridor 8 displays the largest difference between the two scenarios, as a result of its peak traffic characteristics. The two sides of the A10 Motorway Ring are projected to both operate very close to, or just over, capacity at peak. This leads to an unstable situation where small percentage traffic diversions have a major impact on travel speeds. The corridor also suffers the most incidents, and carries the largest traffic volumes over the longest distances included in the study. This means that any differences in travel speeds affect large numbers of motorists, leading to the major cost differences observed in the scenarios.

For corridors where the alternative arterial was small or near to capacity, the result of diverting vehicles from the main route was to improve conditions on the main route but reduce the average speeds on the alternative, thus providing a limited benefit overall. This demonstrates the requirement for realistic alternative routes through a network in order to obtain maximum benefits from a dynamic route guidance system.

The final finding from the analysis of the peak period results relates to main routes with significant spare capacity. The results show that even in situations where there is sufficient spare capacity to accommodate one-lane incidents on the main route, dynamic route guidance can still improve the overall traffic situation significantly by diverting vehicles to realistic alternatives.

Off-peak period analysis

During off-peak period incidents, the benefits of a route guidance system would generally be less than those accrued during peak period incidents. However, the off-peak period is longer than the peak and contains considerably more incidents. Therefore, although the benefit per incident may be small, the overall benefit per year can, in some cases, be significant.

As shown in Tables 7.5 and 7.6, dynamic route guidance would almost always offer some benefit during off-peak period incidents. The maximum benefits obtainable during the off-peak occur when the main route and the alternative arterial have similar characteristics. In these cases, the choice between routes has little consequence until one of the routes is affected by an incident. When this situation arises, a dynamic route guidance system would divert equipped vehicles on to the alternative route, thus providing a benefit to drivers with the system and to those who continue along the affected route.

When flows on the main route are so low that even a type #2 incident does not have any significant effect on the traffic flow, then the effectiveness of a route guidance system for diverting traffic from the main route to an alternative is reduced to a minimum. Even in these cases, however, the modelling results suggest that the overall traffic situation could be improved by the use of dynamic route guidance.

SYSTEM COSTS

This analysis summarises the costs associated with implementing and operating the route guidance systems outlined in the three scenarios for the central Amsterdam region. Costs are incurred in two main areas, the first of which relates to the in-vehicle equipment. This is the only cost area associated with Scenario 1. For Scenarios 2 and 3, the second cost area covers implementation and operation of the route guidance infrastructure. Scenario 3 assumes that some form of traffic monitoring and control system will be available on major routes in the Amsterdam region by the year 2000, and this cost is therefore excluded from the route guidance system evaluation.

Estimates for in-vehicle equipment have been based on information available from system manufacturers. Allowance has been made for expected major reductions in real costs by the year 2000, especially in the mass-produced in-vehicle units. Table 7.7 summarises the costs of the in-vehicle system elements assuming high, medium and low unit costs and associated market penetrations. The market penetrations have been applied to the vehicle population figures supplied by the Rijkswaterstaat to produce total system costs.

Table 7.7   Costs of in-vehicle units for route guidance systems

	Scenario		
	High cost	Medium cost	Low cost
(a) Market penetration (%)	7	14	25
(b) No. of vehicles equipped	44 380	88 760	158 500
SCENARIO 1			
(c) In-vehicle unit cost (DFl)	1200	1000	800
(d) Total cost of in-vehicle unit (DFl million)	53.3	88.8	126.8
SCENARIO 2			
(e) In-vehicle unit cost (DFl)	700	600	500
(f) Total cost of in-vehicle unit (DFl million)	31.1	53.3	79.2
SCENARIO 3			
(g) In-vehicle unit cost (DFl)	1300	1100	900
(h) Total cost of in-vehicle unit (DFl million)	57.7	97.6	142.6
SCENARIO 2A (improved with digital map)			
(i) In-vehicle unit cost (DFl)	1300	1100	900
(j) Total cost of in-vehicle units (DFl million)	57.7	97.6	142.6
SCENARIO 2B (2A + dense beacon network)			
(k) In-vehicle unit cost (DFl)	1300	1100	900
(l) Total cost of in-vehicle units (DFl million)	57.7	97.6	142.6

Scenarios 2A and 2B represent further variants on the basic alternatives in which the low-cost unit of Scenario 2 is replaced by a unit with a digital map, similar to that of Scenario 3.

The implementation and operational costs applicable for Scenarios 2 and 3 can be further divided into two categories; capital and installation costs, and operating costs. Capital and installation costs will be incurred only during the implementation stage of a scheme, or when equipment has reached the end of its operating life and requires replacement. The operating costs will be incurred over the lifetime of the system. These costs can be calculated over a fixed time period, for example annually.

For Scenario 2, the capital and installation costs include:

● purchase and installation of roadside beacons

● purchase and installation of central computer equipment

● purchase and installation of display units and interfaces

● provision of the control centre building

● software development

● coding of network data.

The operating costs for such a system include:

● staff payments

● updating of coded network data

● spare parts and maintenance

● communication and power requirements

● control building overheads.

To calculate the capital and installation costs of roadside beacons, it is first necessary to estimate the number of beacons required to operate successfully an externally linked route guidance system based on infra-red communication. Plans for full implementation of Autoguide in London included 1000 beacons. This gives an average travel time between beacons of approximately 5 minutes, resulting in the 10 minute information lag described in the previous section. Assuming a similar situation in Amsterdam, it is first necessary to scale the number of beacons according to the annual vehicle kilometres travelled in each city. In London, approximately 2900 million vehicle kilometres are travelled per year. In Amsterdam, the approximate annual vehicle kilometres travelled is 700 million. The number of beacons required to provide coverage of Amsterdam can therefore be estimated as 240.

Scenarios 2 and 2A are therefore based on 240 beacons, while Scenario 2B assumes that the beacon spacing is halved, to help reduce the information lag.
The capital and installation cost for each roadside beacon has been estimated at DFl 15 000. Estimates are also available for the costs of software development, network data coding, computer hardware, display units and interfaces, operating staff, and a facility for the central computer.
The cost of coding the network data for Amsterdam has been scaled from published cost data, and is estimated at DFl 1.5 million.
A cost of DFl 1 million has been estimated for computer hardware, DFl 200 000 for display units and interfaces, and DFl 1 million for a facility for the computer equipment and staff. The annual staff cost to operate the system

Table 7.8 Capital, installation and operating costs for an infra-red-based route guidance system

		Cost estimates (DFl)		
		High	Medium	Low
(a)	Capital and installation costs per beacon	20 000	15 000	10 000
(b)	Total capital and installation cost of beacon infrastructure (240 x (a))	4 800 000	3 600 000	2 400 000
(c)	Software development	2 670 000	2 000 000	1 330 000
(d)	Network coding	2 000 000	1 500 000	1 000 000
(e)	Computer hardware	1 330 000	1 000000	670 000
(f)	Display units and interfaces	267 000	200 000	133 000
(g)	Central computer facility	1 330 000	1 000000	670 000
(h)	Communication system hook-up charge per beacon	1 067	800	533
(i)	Total communication system hook-up charge (240 x (h))	256 000	192 000	128 000
(j)	Total implementation cost (b)+(c)+(d)+(e)+(f)+(g)+(i)	12 653 000	9 492 000	6 331 000
(k)	Annual staff costs	670 000	500 000	330 000
(l)	Spare parts and maintenance (15% of (b)+(e)+(f)+(g)+(i))	1 199 000	899 000	600 000
(m)	Monthly communication cost per beacon	94	70	46
(n)	Total annual communication cost (240 x 12 x (m))	271 000	202 000	132 000
(o)	Annual operating cost (k)+(l)+(n)	2 140 000	1 601 000	1 062 000

and to update the network data has been estimated at DFl 500 000. This is based on an estimate of 4 full-time staff, and a number of part-time employees.

The cost of providing communication facilities between the roadside beacons and the central computer depends on the type of system used. Two alternative communication systems were identified.

The first would use a dedicated, voice-grade line with data transmission capability. The approximate installation cost would be DFl 800 per hook-up, with a monthly charge of about DFl 70. The second would be a dial-up service, operating with a modem. The installation cost would be approximately DFl 100, and an environmentally protected modem would cost approximately DFl 400, bringing the total set-up cost to around DFl 500, with a monthly cost of DFl 100. The use of a dedicated line appears preferable; although the set-up cost is higher, the lower monthly charge makes the overall cost lower within a short period.

Since no large-scale implementations of externally linked route guidance exist, reliable data on maintenance and spare parts requirements are not available. These costs have been estimated as 15% of the capital and installation costs of the roadside beacons, computer hardware, display units and interfaces, central computer facility, and the communication system. The capital, installation and operating costs for externally linked route guidance under Scenario 2 are detailed in Table 7.8.

To implement a route guidance system which is updated by RDS-TMC data (Scenario 3), the infrastructure capital and installation cost components are as follows:

• encoders

• central computers.

Running costs are those incurred continuously during the operation of the system. These can therefore be calculated for a fixed time period. The main elements included in the running costs are:

• staff payments

• communication charges

• maintenance.

Implementation of RDS has involved fitting data encoders to existing FM transmitters. Based on these studies and the size of the Amsterdam area, it is anticipated that four encoders would be required. The cost of interfacing with existing encoders is in the range DFl 20 000 to DFl 60 000. High medium and low values have therefore been estimated as DFl 60 000, DFl 40 000 and DFl 20 000.

For a city the size of Amsterdam, only one central computer would be required. The capital and installation cost of the central computer and peripheral equipment has been estimated at approximately DFl 50 000. To allow for variation in this estimate, as with the previous equipment costs, high, medium and low values are assumed. This results in three unit cost scenarios.

The costs outlined above for encoders and the central computer assume that the system is able to interface, or be combined, with existing equipment. This marginal cost analysis, therefore, deals only with additional costs associated with implementation of an RDS-TMC system. It is assumed that the central computer would be located in an existing traffic control centre, and the data encoder would be fitted on existing commercial or public service FM station transmitters. It is also assumed that existing automatic incident detection systems would be used to collect the traffic information for broadcasting via RDS-TMC.

Using these figures for interfacing with the four encoders and for providing one central computer, capital and installation costs have been derived. Table 7.9 shows these results with totals under the high, medium and low cost headings described earlier.

Annual running costs have been calculated in a similar way to the capital and installation costs. High, medium and low cost estimates have been provided. The main costs associated with operating an RDS-TMC system include the labour costs of staff employed to operate the RDS-TMC computer

Table 7.9 Capital, installation and operating costs for an RDS-TMC route guidance system

		Cost estimates (DFl)	
	High	Medium	Low
(a) Capital and installation costs of encoder	60 000	40 000	20 000
(b) Total cost for all encoders (4 x (a))	240 000	160 000	80 000
(c) Central computer facility	60 000	50 000	40 000
(d) Total implementation cost ((b)+(c))	300 000	210 000	120 000
(e) Annual staff costs	144 000	120 000	96 000
(f) Telephone charges	4000	3000	2000
(g) Spare parts and maintenance (10% of (d))	30 000	21 000	12 000
(h) Annual operating cost ((e+(f)+(g))	178 000	144 000	110 000

facility, and the costs of telephone calls between the computer and each of the encoders. Staffing requirements at the control centre have been set at three full-time employees to provide continuous cover. Taking an average salary of DFl 40 000 per year for a trained staff member, the medium cost estimate becomes DFl 120 000. High and low estimates assume a 20% variation in this figure. Telephone charges have been calculated on the same basis as in Scenario 2.

The final element under consideration in the annual running costs is maintenance. In terms of receivers, the maintenance cost is assumed to be negligible. For encoders and the central computer, the annual cost of maintenance and any necessary spare parts has been estimated at 10% of the capital and installation cost.

SUMMARY

Overall summaries of costs and benefits for each of the three main scenarios and three variants are presented in Tables 7.10 to 7.15. The principal scenarios represent implementation of the following three system types in Amsterdam:

(1) route guidance based on self-contained on-board navigation systems;

(2) route guidance based on infra-red communications links to a roadside infrastructure;

(3) route guidance based on communications from RDS-TMC.

The variant scenarios 2A, 2B and 3A examine the effects of including a digital map with the infra-red on-board equipment; increasing the beacon density to reduce the information lag and an alternative assumption of a 10-minute lag with RDS-TMC.

Annual rates of return are calculated by dividing the net annual benefit (i.e. total annual benefits less annual operating costs) by the total capital costs attributed to each scenario.

The savings resulting from semi-dynamic guidance using the autonomous navigation system in Scenario 1 depend on the market penetration achieved by the system. The appraisal of costs for this type of system shows an initial cost (borne by individual users) of the order of twice the annual benefits. This demonstrates that a self-contained navigation system has the potential to pay for itself in two years.

In the case of Scenario 2, the semi-dynamic guidance benefits of an infra-red beacon system are lower than those accruing from the full autonomous navigation system. This is because autonomous guidance would be immediately useful on all journeys, whereas the beacon system would offer

Table 7.10 Summary of quantified benefits and costs for Scenario 1 (self-contained on-board navigation system)

	Scenario		
	High cost	Medium cost	Low cost
Market penetration (%)	7	14	25
Semi-dynamic benefits (DFl million/year)	26	53	94
In-vehicle equipment cost (DFl million)	53.3	88.8	126.8
Annual rate of return	49%	60%	74%

Table 7.11 Summary of quantified benefits and costs for Scenario 2 (infra-red beacon system with 10 minute information lag)

	Scenario		
	High cost	Medium cost	Low cost
Market penetration (%)	7	14	25
Semi-dynamic benefits (DFl million)	22.0	45.0	79.7
Fully dynamic benefits (DFl million)	5.3	12.9	21.1
[Annual operating cost] (DFl million)	(2.1)	(1.6)	(1.1)
Net annual benefit (DFl million)	25.2	56.3	99.7
In-vehicle equipment cost (DFl million)	31.1	53.3	79.2
Infrastructure and installation costs (DFl million)	12.7	9.5	6.3
Total capital cost (DFl million)	43.8	62.8	85.5
Annual rate of return	57%	90%	117%

detailed guidance only after passing a beacon. However, when account is taken of the additional fully dynamic guidance benefits, the net annual benefits of Scenario 2 are similar to those of Scenario 1. As the total capital costs of Scenario 2 are the lowest of all the scenarios, the annual return on investment

Table 7.12 Summary of quantified benefits and costs for Scenario 2A (infra-red beacon system with 10 minute information lag and on-board digital map)

	Scenario		
	High cost	Medium cost	Low cost
Market penetration (%)	7	14	25
Semi-dynamic benefits (DFl million)	27.6	56.2	84.5
Fully dynamic benefits (DFl million)	5.3	12.9	21.1
[Annual operating cost] (DFl million)	(2.1)	(1.6)	(1.1)
Net annual benefit (DFl million)	30.8	67.5	104.5
In-vehicle equipment cost (DFl million)	57.7	97.6	142.6
Infrastructure and installation costs (DFl million)	12.7	9.5	6.3
Total capital cost (DFl million)	70.4	107.1	148.9
Annual rate of return	44%	63%	94%

is significantly greater than that of Scenario 1. This suggests that while both scenarios offer substantial benefits, Scenario 2 should be preferred over Scenario 1.

Inclusion of an on-board digital map with infra-red beacons in Scenario 2A increases the available semi-dynamic benefits. However, on-board equipment costs also rise substantially, causing a fall in the annual rate of return. Some drivers may opt for Scenario 2A in parallel with Scenario 2, if they choose to spend the extra money on the on-board unit. This is probably a matter for market forces rather than for government decision.

Increasing the beacon density (Scenario 2B) similarly increases the fully dynamic benefits available to the infra-red system. However, infrastructure and maintenance costs also rise, further reducing the annual rate of return. This suggests that the investment in the additional beacons may not be worthwhile. Further study is recommended to determine an optimal beacon spacing.

Scenario 3 assumes a full on-board route guidance system and an external link based on RDS-TMC. This option involves the highest total benefits. Costs are lower than those of Scenarios 2A and 2B, due to the very low infra

Table 7.13 Summary of quantified benefits and costs for Scenario 2B (infra-red beacon system with 5 minute information lag and on-board digital map)

	Scenario		
	High cost	Medium cost	Low cost
Market penetration (%)	7	14	25
Semi-dynamic benefits (DFl million)	27.6	56.2	84.5
Fully dynamic benefits (DFl million)	8.1	17.5	28.9
[Annual operating cost] (DFl million)	(3.3)	(2.4)	(1.6)
Net annual benefit (DFl million)	32.4	71.3	111.8
In-vehicle equipment cost (DFl million)	57.7	97.6	142.6
Infrastructure and installation costs (DFl million)	18.4	13.8	9.2
Total capital cost (DFl million)	76.1	111.4	151.8
Annual rate of return	42%	64%	74%

structure and maintenance costs of RDS-TMC. Based on the high to medium cost projections, both Scenarios 2 and 3 offer attractive rates of return. This suggests that, on the basis of the current study, Scenarios 2 and 3 are both worthwhile approaches to the dynamic route guidance problem.

Finally, Scenario 3A examines the effect of a longer information lag with RDS-TMC, perhaps resulting from a less dense or slower incident detection system. The reduction in benefits is small, indicating that the overall result is robust. Rates of return are still very attractive in this, as in all the scenarios.

A particular focus of this case study is the dynamic benefits of route guidance. As has been described in the previous chapter, it is possible to identify corridors in which substantial savings would be achieved using dynamic route guidance during peak and off-peak periods. The corridors in which route guidance equipment would be of major benefit in the event of traffic incidents are generally those in which the following conditions apply:

● Travel demand on the main route is close to normal capacity;

Cost–benefit analysis: a case study

Table 7.14 Summary of quantified benefits and costs for Scenario 3 (RDS-TMC based system with 5 minute information lag)

	Scenario		
	High cost	Medium cost	Low cost
Market penetration (%)	7	14	25
Semi-dynamic benefits (DFl million)	27.6	56.2	84.5
Fully dynamic benefits (DFl million)	8.1	17.5	28.9
[Annual operating cost] (DFl million)	(0.2)	(0.2)	(0.1)
Net annual benefit (DFl million)	35.5	73.5	113.3
In-vehicle equipment cost (DFl million)	57.7	97.6	142.6
Infrastructure and installation costs (DFl million)	0.2	0.2	0.1
Total capital cost (DFl million)	57.9	97.8	142.7
Annual rate of return	61%	75%	79%

- Incidents are of such severity that capacity is significantly reduced;
- The alternative routes have significant reserve capacity.

These factors lead to situations where there is insufficient capacity to accommodate all vehicles on the main route during an incident and there are alternative, less congested routes along which vehicles can be diverted.

The assessment presented here is conservative in that it includes only semi-dynamic benefits, plus dynamic benefits resulting from traffic incidents on major routes. Other traffic situations which could benefit from dynamic route guidance have not been included. Examples include quasi-static situations such as road works lasting several days or weeks; these would be too short to affect long-term traffic routing, yet too long to be covered by the dynamic incident scenarios modelled here. For this reason, the assessment of benefits presented here should be regarded as partial, and the resulting rates of return as conservative estimates.

Table 7.15 Summary of quantified benefits and costs for Scenario 3A (RDS-TMC based system with 10 minute information lag)

	Scenario		
	High cost	Medium cost	Low cost
Market penetration (%)	7	14	25
Semi-dynamic benefits (DFl million)	27.6	56.2	84.5
Fully dynamic benefits (DFl million)	5.3	12.9	21.1
[Annual operating cost] (DFl million)	(0.2)	(0.2)	(0.1)
Net annual benefit (DFl million)	32.7	68.9	105.5
In-vehicle equipment cost (DFl million)	57.7	97.6	142.6
Infrastructure and installation costs (DFl million)	0.2	0.2	0.1
Total capital cost (DFl million)	57.9	97.8	142.7
Annual rate of return	56%	70%	74%

This appraisal of dynamic route guidance for the Amsterdam region can be summarised as follows. Dynamic route guidance would be most effective when the diversion of traffic is sufficient to prevent demand from exceeding capacity and is able to keep demand balanced on all the facilities. These cases are not universal, but are expected to lead to substantial savings.

GLOSSARY

The glossary will both supply a definition for each appropriate term and acronym used in the present course and serve as an index. In addition, a selected number of terms will be "activated" in an attempt to both standardise the definition and to provide the learner with "scaffolding". In this way, an unfamiliar term can be understood by anchoring it to familiar knowledge.

Generic illustration of an activated concept

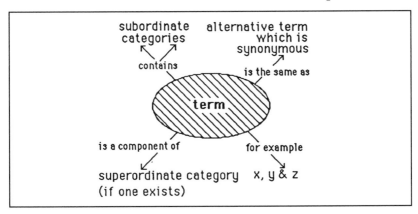

Term or acronym	Definition
AA	The Automobile Association, a UK motoring organisation.
AHAR	Advanced Highway Information Service.
AI	Artificial Intelligence.
ALERT	Advice and Problem Location for European Road Traffic.
algorithm	A computational procedure which uses a defined set of rules to implement a calculation (cf. "heuristic").

Term or acronym	Definition
ALI	Autofahrer Leit und Informationssystem: an externally linked route guidance system developed by Blaupunkt which uses inductive loops for vehicle-roadside communication.
ALI-SCOUT	An externally linked route guidance (infrared based) system (Bosch/Blaupunkt & Siemens), developed from ALI & AUTOSCOUT. Autoguide is the UK development of ALI-SCOUT, now superseded by EURO-SCOUT (Siemens-Plessey).
alphanumeric	A string variable; one from which sensible arithmetic calculations are not possible, because at least some of the component symbols may not be numeric. Also see "numeric".
AMTICS	Advanced Mobile Traffic Information and Communications System. An externally linked route guidance system developed by Japan, and now superceded by the VICS programme.
ANTIOPE	An electronic route planning system (French). This system is not interactive, but collects and displays 30-40 pages of traffic information on a regional basis via a teletext TV service.
application	A function within the field of road transport management which might, potentially, be fulfilled using information technology.

3 Function groups (subdivided into 8 areas) — contains → Application

7 Groups: DRIVE ↗ contains → Application

function — is the same as → Application

is a component of ↙ the IRTE

for example ↘ warning transmission

| ATT | Advanced Road Transport Telematics. This term has largely superseded Road Transport Informatics (RTI) as the European descriptor for the use and operation of road networks through the application of information technology and telecommunications. |

Term or acronym	Definition
Autoguide	An externally linked, automatic route guidance system tried in London. Using infra-red roadside beacons, a central computer provides two-way communication with in-vehicle transceiver units, and thus 'real-time' information. (See also ALI-SCOUT.)
AUTO-SCOUT	An externally linked route guidance system developed by Siemens using infra-red technology for roadside–vehicle communication. Designed to use a sparse network of beacons, with location and route following between beacons achieved by a navigational computer using dead reckoning.
AVI	Automatic vehicle identification.
AVL	Automatic vehicle location.
bandwidth	The range of frequencies used in a specific telecommunications signal.
BAST	Bundes anstalt für Strassenwesen.
baud rate	The rate at which data are transmitted.
beacon	Part of the infrastructure in the physical environment. A roadside device for communicating data to and from vehicles.

beam configuration parameter	Parameters pertaining to the arrangement of beam components.
BBC	British Broadcasting Corporation.

Term or acronym	Definition
Bell Canada	A company which manufactures RTI products.
BER	Bit error rate.
bit	A unit of information.
bit rate	The number of information bits transferred in a given period of time.
BK	The third of three tones which radio broadcasters using ARI are required to utilise in order to provide additional information. The BK tone identifies to which of six predefined areas the messages relate.
Blaupunkt	A company which manufactures RTI products.
BMC	Bakkenist Management Consultants (NL).
Bosch	A company which manufactures RTI products.
bottom-up	Beginning with the most basic components, such as computer hardware or neurones of the brain, and solving a problem using these components rather than with a conceptual or strategical guide. In the context of the present discussion, a bottom-up synthesis commences with the RTI system/technology, then decides in a methodological fashion which RTI application they fulfil.
bit/s	Bits per second.
BT	British Telecom.
burst-mode transmission	This manner of transmission is characterised by periods of transmission and non-transmission which are not necessarily at regular intervals.
cable system	A type of beacon system which allows inter-vehicle and vehicle–roadside communication.
CACS	Comprehensive Automobile Traffic Control System project. An externally linked route guidance system, the prototype of which utilised inductive loop antennas for 2-way communication between vehicle and roadside.

Term or acronym	Definition
CARGOES	A DRIVE project—"Cars and Route Guidance Optimized Electronic Systems"—which is developing integrated traffic control and dynamic route guidance systems.
CARIN	Car Information and Navigation system: an in-vehicle route guidance system.
CARMINAT	A Eureka project involving Renault, Philips and the French broadcasting organisation, TDF. The CARMINAT project devised a communication protocol (P3) concerned with RDS-TMC coding, and generally aims to provide a comprehensive in-vehicle information system.
CCETT	Centre Commun des Etudes Télédiffusion et Télécommunications.
CCIR	International Radio Consultative Committee.
CCITT	International Telephone and Telegraph Consultative Committee.
CD-ROM	Compact Disc-Read Only Memory.
CEC	Commission of the European Communities. The "civil service" of the the European Communities which, among other things, initiates training, research and development programmes such as COMETT, DRIVE, and ESPRIT.
CEPT	European Committee of PTTs.
co-channel interference	Interference from signals on the same frequency, but from an unwanted transmitter.
COMETT	Community Programme for Education and Training for Technology, a CEC initiative. One of the sponsors of *Information Technology on the Move*.
CRC	Castle Rock Consultants (UK).
DACAR	A DRIVE project—"Data Acquisition and Communication Techniques and their Assessment for Road Transport"—with the purpose of reviewing, analysing, developing and demonstrating basic data acquisition and communication techniques for use in RTI systems.

Term or acronym	Definition
dB	Decibels (a logarithmic measure of signal strength).
dBm	Decibels per metre.
DK	The second of three tones which radio broadcasters using ARI are required to utilise in order to provide additional information. The DK tone is added when traffic messages are currently being broadcast.
Doppler effect	A phenomenon observed for sound waves and electromagnetic radiation, characterised by a change in the apparent frequency of a wave as a result of relative motion between the observer and the source. In the case of a microwave system, an energy beam is directed at the road and reflected back by moving objects.
DRG	Dynamic Route Guidance. An RTI application providing in-vehicle information about available and recommended routes based not only on the long-term infrastructure but also on the basis of current and forecasted traffic conditions.
DRIVE	Dedicated Road Infrastructure for Vehicle safety in Europe, a research and development initiative of the CEC, and sponsors of much of the work described in *Information Technology on the Move*.
driver information systems	Provide drivers with information on roadway conditions and route availability.
DTMF	Dual Tone Multi-frequency dialling (as used in modern push-button telephones).
Dutch RVI	Radio Verkeers Informatie. The RVI study (conducted by Philips in 1985) resulted in an early method for coding traffic messages for RDS-TMC.
dynamic	In the context of the present text, this term refers to a system which reflects current changes in the road transport situation.
EBU	European Broadcasting Union.
ECU	European Currency Unit (one unit is approximately £0.7, or equivalent to US$1).
ECMT	European Conference of Ministers of Transport.

Term or acronym	Definition
ERGS	Electronic (externally linked) Route Guidance System. The ERGS concept is based around a two-way communication between vehicles and a roadside beacon network infrastructure via in-pavement inductive loops and vehicle mounted antennas.
ERMES	European Radio Message System: the developing pan-European paging system.
ESPRIT	European Strategic Programme for Research and development in Information Technology—a CEC research programme in IT.
ETSI	European Telecommunications Standard Institute.
ETSI-PS	A study group (previously called RES4) incorporated into ETSI to produce specifications for a paging and messaging system.
Eureka	The European Research Coordination Agency—an initiative in collaborative information technology projects.
EURO-SCOUT	Infrastructure-based route guidance system, part of the Siemens-Plessey "Universal Vehicle Information System". (See also ALI, ALI-SCOUT, Autoguide.)
fade	Loss of radio signal due, for example, to multi-path effects.
fading	A loss of signal strength when information is being transmitted.
"floating cars"	This refers to equipped vehicles who are travelling around the network. Their travel times are used as an input to the route guidance system.
FM	Frequency modulation.
FSK	Frequency-Shift Keying; a means of sending digital information over an analogue communications channel by using one frequency to represent binary "1" and another to represent binary "0".
GHz	Gigahertz (one billion cycles per second).
GPS	Global Positioning System (a satellite-based navigation and location system).
GSC	Galaxy Sequential Code.

Term or acronym	Definition
GSM	Group Special Mobile (the CEPT sub-committee which specified the new Pan-European cellular radio system).
HAR	Highway Advisory Radio (used in the US and Japan to warn of hazards on the road network).
HCI	Human–Computer Interface (sometimes referred to as Human–System Interface, the Man–Machine Interface or the Human–Machine Interface): the way in which information located in an RTI system is revealed to a human. In the present context this human is normally a driver. The term HCI is also used to refer to Human–Computer Interaction, the two-way communication of information between humans and computer interfaces. For additional clarification, refer to "system components" below.
heuristic	A method or set of rules for solving problems other than by a single, fixed, algorithm. A general guiding principle. Computer terminology denoting a rule of thumb.
ICC	Intelligent Cruise Control; an RTI application often referred to as "Co-operative driving".
IDN	Integrated Data Network.
ID-tag	An RTI product (AVI system).
infra-red	A beacon system (used as an alternative to cables), which operates by electromagnetic waves.
infra-structure	Any part of an RTI system that is not located within the vehicle.

sensors;
transmission system;
processing & storage;

contains

Infrastructure

is a component of   for example

The
Physical
Environment
in RTI

road;
roadside;
central computer;
traffic management network

Term or acronym	Definition
intelligence	In the context of the present discussion, this term refers to the degree of information processing occurring and to the use of available information in an adaptive way.
interleaving	A countermeasure used to reduce the bit error probability.
in-vehicle	That part of the RTI system which operates within the physical confines of the vehicle.

An information processing & storage capacity; sensors; HMI

contains

Within the vehicle

is the same as

In-Vehicle

is a component of

The Physical Environment in RTI

IRG	Interactive route guidance.
IRTE	Integrated Road Transport Environment. In antithesis to the currently prevailing "crisis" management, a long-term goal for many in RTI research is the development of an IRTE, whereby there is sufficient information about traffic conditions, use of RTI systems etc., to implement a traffic management plan that makes optimal use of the available road network.
ISDN	Integrated Services Data Network.
ISO	International Organisation for Standardisation; the global top-level standards-making body.
IVHS	Intelligent Vehicle Highway System—U. S. equivalent of RTI or ATT. IVHS America is the Intelligent Vehicle Highway Society of America, a national program which aims to co-ordinate and foster public and private partnerships which contribute to an RTI-based improved transportation system. (See also ATT, RTI, and VICS.)
kbit/s	Kilobits per second.
kHz	Kilohertz; one thousand cycles per second.

Term or acronym	Definition
Kofri	An RTI product (AVI system) manufactured by Micro Design.
LCD	Liquid Crystal Display (could be used, for example, on an HCI display screen).
LISB	A dynamic route guidance system currently being tried in Berlin—"Leit- und InformationsSystem Berlin".
Loran-C	A transmission system utilised by vehicle navigation applications.
Mainframe	Facilities available to users on a given computer network.
Marconi	A company which manufactures RTI products.
Mbit	Megabit.
MCSS	Motorway Control and Signalling System: a road traffic management system commissioned by the Dutch RWS. The infrastructure consists of inductive loop sensors, local roadside processors, and central computers for information storage. A new version of the system will utilise RDS-TMC, for a more dynamic communication process.
MHz	Megahertz; one million cycles per second.
microwave	Electromagnetic radiation in the frequency range 1 GHz to 30 GHz. Frequencies above 30 GHz are generally referred to as millimetric frequencies. Microwave systems have been developed as vehicle sensors (beacons), and offer an alternative to inductive loops and tubes. Microwave systems use a Doppler principle whereby an energy beam is directed at the road, and reflected back off moving objects.
modem	MOdulator-DEModulator; a device which converts digital signals to analogue form for sending along a telephone wire.
modulation	The process of superimposing the frequency of a wave onto another wave (or onto an electron beam). Frequency modulation is a method of transmitting information using a radio-frequency carrier wave.
Motorola	An international company producing electronic components and systems.

Term or acronym	Definition
mu	The 12th letter of the Greek alphabet, represented by the symbol μ.
ms	Milliseconds.
multipath	The phenomenon in (cellular) radio propagation where radio waves reflected from obstacles interfere with the direct waves at the receiver, causing loss of signal (i. e., a "fade").
Navstar GPS	Navstar Global Positioning System. A space-based radio positioning, navigation, and time-transfer system, which should supersede the U.S. Navy TRANSIT system in the early 1990s.
NEC	Nippon Electric Company.
nm	Nanometre (10^{-9}m).
node	The intersection for two or more links. For the purposes of the Route Choice Calculation module, 'node' refers specifically to search types, whereby a route choice problem is represented by a series of nodes and links.
noise	Random effects on a signal that is transmitted, which reduce the probability that the information being transmitted will be correctly received.
numeric	A number which you can do sensible arithmetic calculations with, such as $53 - 3 = 50$.
OECD	Organisation for Economic Cooperation and Development.
OSI	Open Systems Interconnection.
outage probability	The probability that a communication system as a whole does not function, due to transmission errors.
outstation	Part of the infrastructure utilised in a process of decentralisation of information processing.
packet	A discrete block of data within a communication protocol envelope.
PAMELA	A DRIVE project—"Pricing and Monitoring Electronically of Automobiles"—concerned with the design and development of standard equipment for the non-stop debiting of vehicles, with communication between a roadside beacon and a transponder in a vehicle.

Term or acronym	Definition
parameter	A limiting factor.
parity bits	An error-checking approach used in data communications.
P_{eo}	Often used to denote the minimum bit probability error.
Philips	A company which manufactures RTI products.
physical environment	The physical context within which an RTI system will operate.

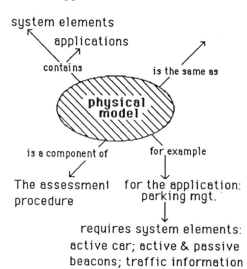

physical model	A non-technical description of the system elements required to fulfil each application.

Term or acronym	Definition
POCSAG	Post Office Code Standardisation Advisory Group. The POCSAG agreed on a code for radio messaging which was adopted as a world standard by the CCIR in 1982 (entitled "CCIR Radio Paging Code number 1: RPCI").
Premid	An RTI product (AVI system).
PRODAT	RTI system for communication between vehicles, particularly a fleet of commercial vehicles.
product	A physical component manufactured to implement an RTI system.

```
technical          synonymous with
components         system in an
            ↖       abstract        ↗
           contains  sense
                          is the same as
              ///////////
           (/// product ///)
              ///////////
                       for example
                              ↘
                    CARIN;
                    MCCS; PRODAT
```

Prometheus	A EUREKA project—"Programme for a European Traffic with Highest Efficiency and unprecedented safety"—undertaken by a consortium of European car manufacturers with the aim of improving driving through the introduction of information technology.
propagation	Signal transmission.
protocols	Sets of rules that structure communication.
PSDN	Public (or packet) Switched Data Network.
PSK	Phase Shift Keying—a modulation technique.
PSTN	Public Switched Telephone Network.
PTTA	Post, Telegraph and Telephone Authority. The standard acronym for the public corporation which runs postal and telephone services in a country.

Term or acronym	Definition
RACS	A Japanese programme called Road and Automobile Communication System, now superseded by the VICS programme.
radio messaging	A digital broadcasting technique used by paging services. It can be defined (CCIR, 1982) as a "non-speech, one-way, personal selective calling system with alert, without message or with a defined message such as numeric or alphanumeric".
radio paging	Another term for radio messaging.
RAM	Random-Access Memory.
RDS	Radio Data System is a subcarrier system which provides a silent data channel on existing VHF-FM radio programmes. A development of RDS is the Traffic Message Channel (TMC).
RDS-TA	Radio Data System—Traffic Announcement.
RDS-TMC	Radio Data System—Traffic Message Channel.
RDS-TP	Radio Data System—Traffic Programme.
RDS transparent data channel	RDS is superimposed onto the normal FM radio signal such that data, in addition to the radio signal, can be broadcast. Only those people with an RDS receiver will be able to decode the transmitted RDS data. As the data does not affect the normal radio broadcast it is considered to be transparent (like teletext on TVs).
ROADWATCH	A service developed by the AA, which provides traffic information to radio stations, television and teletext services.
ROM	Read-Only Memory.
RPCI	CCIR Radio-Paging Code number 1.
RTI	Road Transport Informatics; the application of information technology to road traffic management problems. (See also ATT, IVHS, and VICS.)
Satcom	A general abbreviation for satellite communications.
satisficing	An economic term referring to behaviour which accepts sufficient or non-optimal solutions to problems.
SEL	Standard Elecktrik Lorenz AG; a company which manufactures RTI products.

Term or acronym	Definition
sensor	Any component that extracts information from the physical environment.
shoulder incident	Refers to an incident occurring on the hard shoulder. This is one type of the four incident categories, based on work carried out by Lindley (1987).
Siemens	A company which manufactures RTI products.
SK	The first of three tones which radio broadcasters using ARI are required to utilise in order to provide additional information. The SK tone identifies the radio station as one which regularly carries traffic messages as part of its audio programme.
S/N ratio	Signal-to-noise ratio.
SWAP	System Wide Area Paging; a radio messaging/paging system which was one of the first public, wide-area paging systems, developed in the 1960s.
synergistic	Positively correlated; implies a relationship whereby a growth in one factor leads to a similar growth in the other.
system elements	An RTI system is a means by which an RTI application might be fulfilled. The present text uses this term usually in an abstract sense (whereas the fulfilment of an application technically and physically refers to a product).

vehicles; beacons;
traffic information
&control centres;
broadcasting station.

contains   is the same as

system
elements

is a component of   for example

the physical   a passive,
model   non-responding
vehicle.

| system components | An RTI system can be usefully divided into several component parts. |

Term or acronym	Definition

system components
(*continued*)

TAMS — Telephone Answering Message Service.

TCC — Traffic Control Centres; part of the infrastructure.

techno-phobia — Or techno-fear, refers to a concern about using new and unfamiliar technology.

TELETEL videotext — An RTI product developed in France for pretrip route planning; an electronic route planning system.

TMC — Traffic Message Channel; a development of RDS which will enable traffic messages to be carried digitally and silently without necessarily interrupting the audio programme.

top-down — Top-down analysis begins with the abstract/conceptual or strategical approach rather than from any detailed level. In the present discussion this refers to RTI applications, before deciding which RTI systems potentially fulfil the said application.

transceiver — Transmitter/receiver.

TRANSIT (NNSS) — A US Navy radio positioning system based on four or more satellites in approximately 600 nautical mile polar orbits, together with four ground-based monitoring stations.

TRRL — The UK government Department of Transport's (DTp) Transport and Road Research Laboratory, now known as the TRL (Transport Research Laboratory).

UFFI — User Financing For Infrastructure.

UHF — Ultra-High Frequency. This also refers to the radio frequency band between 300 MHz and 3 GHz.

Term or acronym	Definition
user friendly	This phrase usually refers to certain features of the system/driver interface which make the system more acceptable and more usable to the human operator. Recent developments have resulted in certain features which potentially increase the degree to which an interface is user-friendly (such as voice recognition). User friendly refers to characteristics such as simplicity, accessibility, and familiarity.
UVIS	Universal Vehicle Information System, developed by Siemens-Plessey. An information system for collecting traffic data (parking, public transport, road tolls, tourist information etc.) from a number of sources, including road loops and roadside beacons, and delivery to the vehicle through the EURO-SCOUT and RDS-TMC systems.
VALEO	A car components manufacturer working within the DACAR project, and involved in field trials of infrared communications between vehicles and roadside infrastructure.
VHF	Very High Frequency. This also refers to the radio frequency band between 30 and 300 MHz.
VICS	Vehicle Information and Communication System—the Japanese initiative to introduce RTI, developed from the RACS and AMTICS programmes. (See also RTI, ATT and IVHS.)
watt	A unit of power. It is the derived SI unit of power, equal to 1 joule per second; the power dissipated by a current of 1 ampere flowing across a potential difference of 1 volt. 1 watt is equivalent to 1.341×10^{-3} horsepower (symbol: W).
waveguide	A device that can be used for guiding microwaves.
wave resistor	An electrical component designed to introduce a known value of resistance into a circuit.

REFERENCES

Aicher, P., Busch, F., and Gloger, R., 1991. Improvements to traffic control systems and policies by means of interactive route guidance data. Part A: Use of route guidance data for better traffic control and data management in integrated systems *Advanced Telematics in Road Transport: Proceedings of the DRIVE Conference, Brussels* Elsevier: Amsterdam.

Altendorf, J., Andrisano, O., van Berkel, P., and Frullone, M., 1990. An assessment scheme for transmission systems applied to road transport information *Proceedings of the SAE Future Transportation Technology Conference (SAE Technical Paper Series)* San Diego.

Altendorf, J., Andrisano, O., van Berkel, P., Falciasecca, G., Form, P., Frullone, M., Hengeveld, W., and Immovilli, G., 1991. Assessment methodology and criteria for RTI transmission systems *Advanced Telematics in Road Transport: Proceedings of the DRIVE Conference, Brussels* Elsevier: Amsterdam.

Beccaria, G., Bruschieri, P., Lanteri, F., and Alamia, G., 1991. Improvements to traffic control systems and policies by means of interactive route guidance data. Part B: An integrated observer for the intersection—use of dynamic route guidance data for improving the real time estimation *Advanced Telematics in Road Transport: Proceedings of the DRIVE Conference, Brussels* Elsevier: Amsterdam.

Blythe, P. T., Korolkiewicz, E., Givens, J., Dadds, A. F., Stocker, B. J., and Holford, K., 1991. A short-range road to vehicle microwave communications link for automatic debiting and other RTI services *Advanced Telematics in Road Transport: Proceedings of the DRIVE Conference, Brussels* Elsevier: Amsterdam.

Boheim, M. and Fischer, H. J., 1991. A 61 GHz link: potentiality, feasibility, results *Advanced Telematics in Road Transport: Proceedings of the DRIVE Conference, Brussels* Elsevier: Amsterdam.

Bolelli, A., Mauro, V., and Perono, E., 1991a. Models and strategies for dynamic route guidance. Part B: A decentralised, fully dynamic, infrastructure supported route guidance *Advanced Telematics in Road Transport: Proceedings of the DRIVE Conference, Brussels* Elsevier: Amsterdam.

Bolelli, A., Mauro, V., and Perono, E., 1991b. Improvements to traffic control systems and policies by means of interactive route guidance data. Part C: The combined use of conventional and not conventional RTI systems measurements for the dynamic O/D and traffic pattern estimation in urban networks *Advanced Telematics in Road Transport: Proceedings of the DRIVE Conference, Brussels* Elsevier: Amsterdam.

Bonsall, P. and Bell, M., 1987. *Information Technology: Applications in Transport* VNU Science Press: Utrecht, Netherlands.

Bonsall, P., Pickup, L., and Stathopoulos, A., 1991. Measuring behavioural responses to road transport informatics *Advanced Telematics in Road Transport: Proceedings of the DRIVE Conference, Brussels* Elsevier: Amsterdam.

Bright, J. and Ayland, N., 1991. Evaluating real-time responses to in-vehicle driver information systems *Advanced Telematics in Road Transport: Proceedings of the DRIVE Conference, Brussels* Elsevier: Amsterdam.

Briscoe, E. J., 1987. *Modelling Human Speech Comprehension: A Computational Approach* Ellis Horwood: Chichester.

Bueno, S. and Ongaro, D., 1991. Vehicle/roadside communication and route guidance *Advanced Telematics in Road Transport: Proceedings of the DRIVE Conference, Brussels* Elsevier: Amsterdam.

Catling, I., 1987 The London Autoguide demonstration scheme. Paper presented at the *PTRC Seminar on Information Technology in Transport and Tourism* University of Bath.

Catling, I., Op de Beek, F., Casimir, C., Mannings, R., Zijderhand, F., Zechnall, W., and Hellaker, J., 1991. SOCRATES: System of cellular radio for traffic efficiency and safety. *Advanced Telematics in Road Transport: Proceedings of the DRIVE Conference, Brussels* Elsevier: Amsterdam.

CCIR, 1982. *Recommendations and Reports of the CCIR*, VIII. ITU: Geneva.

Charbonnier, C., Farges, J. L., and Henry, J. J., 1991. Models and strategies for dynamic route guidance. Part C: Optimal control approach *Advanced Telematics in Road Transport: Proceedings of the DRIVE Conference, Brussels* Elsevier: Amsterdam.

Cheung, Y.H.F., Kleijn, H.J. and Gunn, H.F., 1989. *The Netherlands Value-of-travel-time Study: Results and Policy Implications*, Hague Consulting Group and Ministry of Transport and Public Works, The Netherlands.

Davies, P. and Klein, G., 1991. RDS-ALERT—Advice and problem location for European road traffic *Advanced Telematics in Road Transport: Proceedings of the DRIVE Conference, Brussels* Elsevier: Amsterdam.

Dijkstra, E. W., 1959. A note on two problems in connection with graphs *Numerische Mathematics* **11**, 269–271.

Ely, S. R and Jeffrey, D. J., 1990. Traffic Information Broadcasting and RDS. In Walker, J. (ed.), *Mobile Information Systems*, Artech: Boston, MA.

Ernst, H., 1991. The DACAR inductive cable for communication and vehicle control *Communication Techniques in Road Traffic Informatics: Proceedings of the DACAR Conference, Rome and Brussels* Bakkenist Management Consultants: Amsterdam.

Evans, L., 1985. Human Behaviour Feedback and Traffic Safety *Human Factors*, **27**, 555–576.

French, R. L., 1986. Automobile navigation: where is it going? *Proceedings of IEEE Position Location and Navigation Symposium* Las Vegas, NV.

French, R. L. and Lang, G. M., 1973 Automatic route control system *IEEE Transactions on Vehicular Technology*, **22** (2).

Fuller, R., 1984. A conceptualisation of driving behaviour as threat avoidance *Ergonomics*, **27**, 1139–1155.

Fuller, R., 1988. On learning to make risky decisions *Ergonomics*, **31**, 519–526.

Giesa, S. and Everts, K., 1987. ARIAM: Car-driver radio information on the basis of automatic incident detection *Traffic Engineering and Control* (June), 334–348.

Gould, J. D., 1988. How to design systems. In Helander, M. (ed.) *Handbook of Human–Computer Interaction* Elsevier: Amsterdam.

Greenstein, J. S. and Arnaut, L. Y., 1988. Input devices. In: Helander, M. (ed.) *Handbook of Human–Computer Interaction* Elsevier: Amsterdam.

Hakkert, A.S. and Mahahel, D., 1978. The effect of traffic signals on road accidents with particular reference to the introduction of a blinking green phase *Traffic Engineering and Control*, **19** (5), 212–215.

Hart, P. E., Nilsson, N. J., and Raphael, B., 1968. A formal basis for the heuristic determination of minimum cost paths *IEEE Transactions on Systems Science and Cybernetics*, **4**, 100–107.

Helander, M., 1988. *Handbook of Human–Computer Interaction* Elsevier: Amsterdam.

Hills, P. J. and Blythe, P. T., 1990. A system for the non-stop automatic debiting of vehicles *Proceedings of the 22nd ISATA*, Florence.

Höfgen, G., Moss, C. R., and Wörz, T., 1991. The 60 GHz short range link *Communication Techniques in Road Traffic Informatics: Proceedings of the DACAR Conference, Rome and Brussels* Bakkenist Management Consultants: Amsterdam.

Hounsell, N., McDonald, M., and Breheret, L., 1991. Models and strategies for dynamic route guidance. Part A: The modelling of dynamic route guidance *Advanced Telematics in Road Transport: Proceedings of the DRIVE Conference, Brussels* Elsevier: Amsterdam.

Howarth, C. I., 1987. Perceived risk and behavioural feedback: strategies for reducing accidents and increasing efficiency *Work & Stress*, **1**, 61–65.

Howarth, C. I., 1988. The relationship between objective risk, subjective risk and behaviour *Ergonomics*, **31**, 527–535.

Janssen, W. and Tenkink, E., 1988. Risk homeostasis theory and its critics: Time for an agreement *Ergonomics*, **31**, 429–433.

Jeffrey, D. J., 1981. The potential benefits of route guidance. Transport and Road Research Laboratory (TRRL). Laboratory Report 997, Crowthorne.

Jeffrey, D. J., 1985. Options for the provision of improved driver information systems: the role of microelectronics and information technology *IEE Digest No. 1985/11*, London.

Jeffrey, D. J., 1987. Route guidance. In: Bonsall, P. and Bell, M. (eds.) *Information Technology Applications in Transport* VNU Science Press: Utrecht, Netherlands.

Jeffrey, D. J., Russam, K., and Robertson, D. I., 1987. Electronic route guidance by Autoguide *Proceedings of PTRC Seminar on Information Technology in Transport and Tourism* University of Bath.

Jiang, C., Underwood, G., and Howarth, C.I., 1992. Towards a theoretical model for behavioural adaptations to changes in the road transport system *Transport Reviews*, **12**, 253–264.

Klebelsberg, D., 1977. Das Moldell der subjektiven und objektiven Sicherheit *Schweizerische Zeitschrift für Psychologie und ihre Anwendungen*, **36**, 285–294.

Lindley, J.A., 1987. A methodology for quantifying urban freeway congestion. Paper presented at the Transportation Research Board, Washington DC.

Marco, L. and Flowerdew, A., 1991. PARIS - Project for the economic assessment of RTI systems *Advanced Telematics in Road Transport: Proceedings of the DRIVE Conference, Brussels* Elsevier: Amsterdam.

Näätänen, R. and Summala, H., 1974. A model for the role of motivational factors in drivers' decision-making *Accident Analysis and Prevention*, **6**, 243–261.

Nicholson, T. A. J., 1966. Finding the shortest route between two points in a network *Computer Journal*, **9**, 275–280.

Nilsson, N. J., 1980. *Principles of Artificial Intelligence*. Tioga: Palo Alto, CA.

OECD, 1990. *Road Transport Research—Behavioural Adaptations to Changes in the Road Transport System* OECD Publications: Paris.

O'Neill, B., 1977. A decision-theory model of danger compensation *Accident Analysis and Prevention*, **10**, 157–165.

Ott, G. D., 1977. Vehicle location in cellular mobile radio systems *IEE Transactions on Vehicular Technology*, **26**(1), 43–60.

Parkes, A. M., 1991. Data capture techniques for RTI usability evaluation *Advanced Telematics in Road Transport: Proceedings of the DRIVE Conference, Brussels* Elsevier: Amsterdam.

Pearl, J., 1984. *Heuristics: Intelligent Search Strategies for Computer Problem Solving* Addison-Wesley, Reading, MA.

Reichman, R., 1985. *Getting Computers to Talk Like You and Me* MIT Press: Cambridge, MA.

Rumar, K., Berggrund, U., Jerberg, P. and Ytterbom, U., 1976. Driver reaction to a technical safety measure - studded tyres *Human Factors*, **18**, 443–454.

Schneider, H. W., 1985. ARIAM—a car-driver radio information system with automatic text arrangement and display *Proceedings of International Seminar on Electronics and Traffic on Major Roads—Technical, Reglementary and Ergonomic Aspects* Commission of the European Communities: Paris.

Shields, T. R., 1991. Program and technology of IVHS-America *Communication Techniques in Road Traffic Informatics: Proceedings of the DACAR Conference, Rome and Brussels* Bakkenist Management Consultants: Amsterdam.

Shneiderman, B., 1992. *Designing the User Interface: Strategies for Effective Human-Computer Interaction* Addison-Wesley: Reading, MA.

Snyder, H. L., 1988. Image quality. In: Helander, M. (ed.) *Handbook of Human–Computer Interaction* Elsevier: Amsterdam.

Streeter, L. A., 1988. Applying speech synthesis to user interfaces. In: Helander, M. (ed.) *Handbook of Human–Computer Interaction* Elsevier: Amsterdam.

Thoone, M. L. G., 1987. CARIN: A car information and navigation system *Philips Technical Review*, December.

Tridgell, R. H., 1987. Experience of CCIR radiopaging code no. 1 *ITU Telecommunication Journal*, **54**(3).

Tridgell, R. H., 1990. Radiopaging and messaging. In: Walker, J. (ed.) *Mobile Information Systems* Artech: Boston, MA.

Tullis, T. S., 1988. Screen design. In: Helander, M. (ed.) *Handbook of Human–Computer Interaction* Elsevier: Amsterdam.

Underwood, G., 1982. Attention and awareness in cognitive and motor skill, In: Underwood, G. (ed.) *Aspects of Consciousness*, Vol.3, Academic Press: London.

Underwood, G., Jiang, C., and Howarth, C. I., 1993. Modelling of safety measure effects and risk compensation *Accident Analysis and Prevention*, **25**.

Underwood, J. D. M. and Underwood, G., 1990. *Computers and Learning* Blackwell: Oxford.

van der Hart, L. H. M., Toffano, J., and Boucheron, J. L., 1991. The DACAR infrared communication systems *Advanced Telematics in Road Transport: Proceedings of the DRIVE Conference, Brussels* Elsevier: Amsterdam.

van der Molen, H. H. van der and Bötticher, A. M. T., 1988. A hierarchical risk model for traffic participants *Ergonomics*, **31**, 537–556.

von Tomkewitsch, R., 1987. LISB: large-scale test "navigation and information system Berlin" *Proceedings of PTRC Seminar on Information Technology in Transport and Tourism* University of Bath.

von Tomkewitsch, R. & Kossack, J., 1989. *IRTE Argumentation Guide* Siemens: Munich.

Walker, J., 1990. *Mobile Information Systems* Artech: Boston, MA.

Wall, N. D. C. and Williams, D. H., 1991. DRIVE integrated communications infrastructure *Advanced Telematics in Road Transport: Proceedings of the DRIVE Conference, Brussels* Elsevier: Amsterdam.

Wilde, D. and Beightler, C. S., 1967. *Foundations of Optimization,* Prentice-Hall: Englewood Cliffs, NJ.

Wilde, G., 1982. The theory of risk homeostasis: implications for safety and health *Risk Analysis,* **2**, 209–225.

Wilde, G., 1988. Risk homeostasis theory and traffic accidents: propositions, deductions and discussion of recent commentaries *Ergonomics,* **31**, 441–468.

Wilson, W. and Anderson, J., 1980. The effects of tyre type on driving speed and presumed risk taking *Ergonomics,* **23**, 223–235.

Wootton, J. and Ness, M., 1989. The experience of developing and providing driver route information systems. Paper presented to the Institution of Electrical Engineers, London.

AUTHOR INDEX

SUBJECT INDEX